Solitary Waves in Fluids

WITPRESS

WIT Press publishes leading books in Science and Technology.
Visit our website for the current list of titles.
www.witpress.com

WITeLibrary

Home of the Transactions of the Wessex Institute, the WIT electronic-library provides the
international scientific community with immediate and permanent access to individual
papers presented at WIT conferences. Visit the eLibrary at
http://library.witpress.com

International Series on Advances in Fluid Mechanics

Objectives

The field of fluid mechanics is rich in exceptional researchers worldwide who have advanced the science and brought a greater technical understanding of the subject to their institutions, colleagues and students.

This book series has been established to bring such advances to the attention of the broad international community. Its aims are achieved by contributions to volumes from leading researchers by invitation only. This is backed by an illustrious Editorial Board who represent much of the active research in fluid mechanics worldwide.

Volumes in the series cover areas of current interest and active research and will include contributions by leaders in the field.

Topics for the series include: Bio-Fluid Mechanics, Biophysics and Chemical Physics, Computational Methods for Fluids, Experimental & Theoretical Fluid Mechanics, Fluids with Solids in Suspension, Fluid-Structure Interaction, Geophysics, Groundwater Flow, Heat and Mass Transfer, Hydrodynamics, Hydronautics, Magnetohydrodynamics, Marine Engineering, Material Sciences, Meteorology, Ocean Engineering, Physical Oceanography, Potential Flow of Fluids, River and Lakes Hydrodynamics, Slow Viscous Fluids, Stratified Fluids, High Performance Computing in Fluid Mechanics, Tidal Dynamics, Viscous Fluids, and Wave Propagation and Scattering.

Series Editor

M. Rahman
DalTech, Dalhousie University, Halifax,
Nova Scotia, Canada

Assistant Series Editor

M.G. Satish
DalTech, Dalhousie University, Halifax,
Nova Scotia, Canada

T. Matsui
Nagoya University
Japan

A.C. Mendes
Universidade de Beira Interior
Portugal

T.B. Moodie
University of Alberta
Canada

M. Ohkusu
Kyushu University
Japan

E. Outa
Waseda University
Japan

W. Perrie
Bedford Institute of Oceanography
Canada

H. Pina
Instituto Superior Tecnico
Portugal

H. Power
University of Nottingham
UK

D. Prandle
Proudman Oceanographic Laboratory
UK

K.R. Rajagopal
Texas A & M University
USA

D.N. Riahi
University of Illinois-Urbana
USA

P. Škerget
University of Maribor
Slovenia

G.E. Swaters
University of Alberta
Canada

P.A. Tyvand
Agricultural University of Norway
Norway

R. Verhoeven
Ghent University
Belgium

M. Zamir
University of Western Ontario
Canada

Solitary Waves in Fluids

Editor

R. H. J. Grimshaw
Loughborough University, UK

WITPRESS Southampton, Boston

Editor:
R. H. J. Grimshaw
Loughborough University, UK

Published by

WIT Press
Ashurst Lodge, Ashurst, Southampton, SO40 7AA, UK
Tel: 44 (0) 238 029 3223; Fax: 44 (0) 238 029 2853
E-Mail: witpress@witpress.com
http://www.witpress.com

For USA, Canada and Mexico

WIT Press
25 Bridge Street, Billerica, MA 01821, USA
Tel: 978 667 5841; Fax: 978 667 7582
E-Mail: infousa@witpress.com
http://www.witpress.com

British Library Cataloguing-in-Publication Data

A Catalogue record for this book is available
from the British Library

ISBN-13: 978-1-845641-57-3
ISSN: 1353-808X

LOC: 2006938301

The text of the chapters of this book were set individually by the
authors and under their supervision.

Contents

Preface

The history of the solitary wave begins with the well-known observations and experiments of John Scott Russell, which he reported in detail in 1844. Initially controversial, his work was confirmed by the theoretical work of Boussinesq and Rayleigh in the 1870s followed in 1895 by the famous paper by Korteweg and de Vries, describing the equation that now bears their names. Interest in solitary waves subsided until the 1960s when the seminal numerical study of the Korteweg-de Vries equation by Kruskal and Zabusky was rapidly followed by the discovery of the integrability of that equation and also of several other key nonlinear wave equations such as the nonlinear Schrödinger equation, together with the central role of their soliton solutions. The modern theory of solitons and integrable systems has developed from this basis to become a major mathematical field in its own right. At the same time, it became apparent that the theory of solitons had applications in many physical areas, including fluids in a variety of different settings.

In this book, we describe how the theory of solitary waves (i.e. solitons) has been exploited in several fluid flow contexts. Our emphasis is primarily on flows in a geophysical and environmental framework, but the basic methodologies can clearly be used in many other fluid settings. The Introduction (Chapter 1) by Roger Grimshaw sets out the two basic paradigms for solitary waves in fluids. Both cases can be described as being due to a bifurcation in wavenumber space from those points in the linear spectrum where the phase and group velocities are equal. For long waves, this bifurcation takes place at zero wavenumber, and for weakly nonlinear waves leads to the Korteweg-de Vries equation as the appropriate model. The other class of solitary waves arises due to a bifurcation at a finite, non-zero wavenumber, and for weakly nonlinear waves leads to the nonlinear Schrödinger equation for the wave envelope, and hence to the concept of envelope solitary waves.

Then in Chapter 2, Gennady El presents the modern theory of the Korteweg-de Vries equation based on the inverse scattering transform and Whitham modulation theory, and how this is used to describe solitons and undular bores. In Chapter 3 Jean-Marc Vanden-Broeck describes how numerical methods are being used to provide a comprehensive description of solitary waves on water, including the effects of surface tension, and covering an amplitude range from weakly nonlinear waves to large-amplitude waves close to the limiting case. Chapter 4 by Efim Pelinovsky, Oxana Polukhina, Alexey Slunyaev and Tatiana Talipova describes how various models based on the Korteweg-de Vries equation have been used to study internal solitary waves in the ocean. In Chapter 5 Triantaphyllos Aklyas describes the effect of background rotation on solitary waves; in one case, background swirl provides a waveguide for solitary waves; in another case rotation provides a mechanism for solitary waves to decay through radiation. In Chapter 6, John Boyd discusses the various kinds of solitary waves that occur on the planetary scale, where the earth's rotation is a dominant effect. Finally, in Chapter 7 Roger Grimshaw presents an outline of the theory of envelope solitary waves, and the implications of that theory for gravity-capillary waves.

The Editor
Loughborough, 2006

CHAPTER 1

Introduction

R.H.J. Grimshaw
Department of Mathematical Sciences, Loughborough University,
Loughborough, UK.

Abstract

Two distinct paradigms exist for solitary waves in fluids. Both kinds arise due to a bifurcation in wavenumber space from those points in the linear spectrum where the phase and group velocities are equal. For long waves, this bifurcation takes place at zero wavenumber, and for weakly nonlinear waves leads to the well-known Korteweg–de Vries equation and its soliton solutions. We present a brief historical discussion and implications for solitary waves in fluids. The other class of solitary waves arises due to a bifurcation at a finite, non-zero wavenumber, and for weakly nonlinear waves leads to the nonlinear Schrödinger equation for the wave envelope. The soliton solution then provides a description of an envelope solitary wave. In this introductory chapter, we shall give a brief historical account of the Korteweg–de Vries equation and its role in describing solitary waves in fluids, followed by an analogous account of envelope solitary waves.

1 Historical introduction

The Korteweg–de Vries (KdV) equation, given here in canonical form,

$$A_t + 6AA_x + A_{xxx} = 0, \tag{1}$$

is widely recognized as a paradigm for the description of weakly nonlinear long waves in many branches of physics and engineering. It has been established as a relevant model for many fluid flow situations. Here, $A(x, t)$ is an appropriate field variable describing the wave amplitude, t is the time, and x is the space coordinate in the propagation direction. It describes how waves evolve under the competing but comparable effects of weak nonlinearity and weak dispersion. Indeed, if it is supposed that x-derivatives scale as ϵ, where ϵ is the small parameter characterizing long waves (i.e. typically the ratio of a relevant background length scale to a wavelength scale), then the amplitude scales as ϵ^2 and the time evolution, relative to a frame of reference propagating with the relevant linear long-wave speed, takes place on a scale of ϵ^{-3}. The KdV equation is characterized by its family of solitary wave solutions,

$$A = a\mathrm{sech}^2(\gamma(x - Vt)), \quad \text{where } V = 2a = 4\gamma^2. \tag{2}$$

This solution describes a single-parameter family of steady isolated wave pulses of positive polarity; a convenient choice of parameter is the wavenumber γ; note that the speed V is proportional to the wave amplitude a and to the square of the wavenumber γ^2.

The KdV equation (1) owes its name to the famous paper of Korteweg and de Vries, published in 1895, in which they showed that small-amplitude long waves on the free surface of water could be described by the equation

$$\zeta_t + c\zeta_x + \frac{3c}{2h}\zeta\zeta_x + \frac{ch^2}{6}\delta\zeta_{xxx} = 0. \tag{3}$$

Here $\zeta(x, t)$ is the elevation of the free surface relative to the undisturbed depth h, $c = (gh)^{1/2}$ is the linear long-wave phase speed and $\delta = 1 - 3B$, where $B = \sigma/gh^2$ is the Bond number measuring the effects of surface tension ($\rho\sigma$ is the coefficient of surface tension and ρ is the water density). The transformation to a reference frame moving with the speed c (i.e. (x, t) is replaced by $(x - ct, t)$) and subsequent rescaling readily establish the equivalence of (1) and (3). Although equation (1) now bears the name KdV, it was apparently first obtained by Boussinesq (1877) (see Miles 1980; Pego and Weinstein 1997; Nekorkin and Velarde 2002 for historical discussions on the KdV equation). Korteweg and de Vries found the solitary wave solutions (2) and, importantly, showed that they are the limiting members of a three-parameter family of periodic travelling wave solutions of (1), described by elliptic functions and commonly called cnoidal waves,

$$A = d + a\mathrm{cn}^2(\gamma(x - Vt)|m), \tag{4}$$

where

$$V = 6d + 4(2m - 1)\gamma^2 \quad \text{and} \quad a = 2m\gamma^2. \tag{5}$$

Here $\mathrm{cn}(x|m)$ is the Jacobian elliptic function of modulus m ($0 < m < 1$). As $m \to 1$, $\mathrm{cn}(x|m) \to \mathrm{sech}(x)$ and then the cnoidal wave (4) becomes the solitary wave (2), now riding on a background level d. On the other hand, as $m \to 0$, $\mathrm{cn}(x|m) \to \cos 2x$ and so the cnoidal wave (4) collapses to a linear sinusoidal wave (note that in this limit $a \to 0$).

This solitary wave solution found by Korteweg and de Vries had earlier been obtained directly from the governing equations (in the absence of surface tension) independently by Boussinesq (1871, 1877) and Rayleigh (1876), who were motivated to explain the now very well-known observations and experiments of Russell (1844). Curiously, it was not until quite recently that it was recognized that the KdV equation is not strictly valid if surface tension is taken into account and $0 < B < 1/3$, as then there is a resonance between the solitary wave and very short capillary waves (see Section 3).

After this ground-breaking work of Korteweg and de Vries, interest in solitary water waves and the KdV equation declined until the dramatic discovery of the *soliton* by Zabusky and Kruskal in 1965. Through numerical integrations of the KdV equation, they demonstrated that the solitary wave (2) could be generated from quite general initial conditions and could survive intact collisions with other solitary waves, leading them to coin the term soliton. Their remarkable discovery, followed almost immediately by the theoretical work of Gardner *et al.* (1967) showing that the KdV equation was *integrable* through an inverse scattering transform, led to many other startling discoveries and marked the birth of soliton theory as we know it today. The implication is that the solitary wave is the key component needed to describe the behaviour of long, weakly nonlinear waves. These aspects are fully explored in Chapter 2.

The KdV equation (1) is unidirectional. A two-dimensional version of the KdV equation is the Kadomtsev and Petviashvili (KP) equation (Kadomtsev and Petviashvili 1970),

$$(A_t + 6AA_x + A_{xxx})_x \pm A_{yy} = 0. \qquad (6)$$

This equation includes the effects of weak diffraction in the y-direction, in that y-derivatives scale as ϵ^2 whereas x-derivatives scale as ϵ. Like the KdV equation it is an integrable equation. When the '+' sign holds in (6) (called the KPII equation), it can be shown that then the solitary wave (2) is stable to transverse disturbances. This is the case for water waves when $0 < B < 1/3$. On the other hand if the '−' sign holds (called the KPI equation), the solitary wave is unstable; instead this equation supports 'lump' solitons. This is the case for water waves when $B > 1/3$. Both KPI and KPII are integrable equations. To take account of stronger transverse effects, and/or to allow for bidirectional propagation in the x-direction, it is customary to replace the KdV equation with a Boussinesq system of equations; these combine the long-wave approximation to the dispersion relation with the leading order nonlinear terms and occur in several asymptotically equivalent forms.

Although the KdV equation (1) is historically associated with water waves, it also occurs in many other physical contexts, where it arises as an asymptotic multiscale reduction from the relevant governing equations. Typically the outcome is

$$A_t + cA_x + \mu AA_x + \delta A_{xxx} = 0. \qquad (7)$$

Here, c is the relevant linear long-wave speed for the mode whose amplitude is $A(x, t)$, while μ and δ, the coefficients of the quadratic nonlinear and linear dispersive terms, respectively, are determined from the properties of this same linear long-wave mode and, like c, depend on the particular physical system being considered. Note that the linearization of (7) has the linear dispersion relation

$\omega = ck - \lambda k^3$ for linear sinusoidal waves of frequency ω and wavenumber k; this expression is just the truncation of the full dispersion relation for the wave mode being considered and immediately identifies the origin of the coefficient δ. Similarly, the coefficient μ can be identified with an amplitude-dependent correction to the linear wave speed. The transformation to a reference frame moving with a speed c and subsequent rescaling show that (7) can be transformed to the canonical form (1). Equations of the form (7) arise in the study of internal solitary waves in the atmosphere and ocean, mid-latitude and equatorial planetary waves, plasma waves, ion–acoustic waves, lattice waves, waves in elastic rods and in many other physical contexts (see, for instance, Ablowitz and Segur 1981; Dodd *et al.* 1982; Drazin and Johnson 1989; Grimshaw 2001, 2005 as well as Chapters 4, 5 and 6).

2 Korteweg–de Vries equation

2.1 Derivation using a multiscale asymptotic expansion

For simplicity, we shall describe the derivation for the case of gravity waves on the free surface of water, in the absence of surface tension. Let us consider a one-dimensional wave field so that the free surface is represented by $z = \zeta(x, t)$ for an incompressible inviscid fluid with constant density ρ, occupying the region $-h < z < \zeta$, where the undisturbed depth is h. The velocity field is $\mathbf{u} = (u, w)$ and can be assumed to be irrotational, so that $\mathbf{u} = \nabla\phi$, where $\phi(x, z, t)$ satisfies Laplace's equation in $(x . z)$; at the rigid bottom, $w = 0$ at $z = -h$. At the free surface, the flow must satisfy two conditions

$$\zeta_t + u\zeta_x = w \quad \text{at } z = \zeta, \tag{8}$$

$$\phi_t + \frac{|\mathbf{u}|^2}{2} + g\zeta = 0 \quad \text{at } z = \zeta. \tag{9}$$

The first of these is the kinematic condition and the second is the condition for constant pressure, where a Bernoulli relation has been used.

There are now several methods that can be used to extract a reduction to the KdV equation, but here we shall use a multiscale asymptotic expansion which is a versatile approach and can be adapted to many other situations. Thus, we introduce two small parameters, $\epsilon \ll 1$ measuring linear wave dispersion and $\alpha \ll 1$ measuring nonlinearity. More formally, we can define $\epsilon = h/\lambda$, where λ is a typical wavelength of the disturbance, and $\alpha = a/h$, where a is a typical wave amplitude. We then rescale the horizontal coordinate and the time,

$$X = \epsilon x, \quad T = \epsilon t, \tag{10}$$

and seek an asymptotic expansion of the form

$$\zeta = \alpha\zeta^{(0)}(X, T) + \alpha^2\zeta^{(1)}(X, T) + \cdots . \tag{11}$$

There is a similar expansion for the other fluid variables. For this water wave problem, it is convenient to define the depth-averaged mean flow

$$U(X,T) = \frac{1}{h+\zeta} \int_{-h}^{\zeta} u(X,T,z)\,dz. \tag{12}$$

Then it is readily shown that the conservation of mass implies that

$$\zeta_T + (U(h+\zeta))_X = 0. \tag{13}$$

Then at leading order we get

$$\zeta_T^{(0)} + hU_X^{(0)} = 0, \tag{14}$$

$$U_T^{(0)} + g\zeta_X^{(0)} = 0. \tag{15}$$

This is readily reduced to the one-dimensional D'Alembert wave equation, whose general solution is a superposition of a wave propagating in the positive x-direction with speed c and a wave propagating in the negative x-direction also with speed c, where we recall that $c = (gh)^{1/2}$. We choose a wave propagating to the right, so that

$$\zeta^{(0)} = \frac{h}{c}U^{(0)} = A(\xi,\tau), \tag{16}$$

where

$$\xi = X - cT \quad \text{and} \quad \tau = \alpha T. \tag{17}$$

Here, we have anticipated the need for a longer time scale through the slow variable τ to take account of the effects of weak nonlinearity and weak dispersion. Relative to the leading order terms, these are $O(\alpha)$ and $O(\epsilon^2)$ respectively. Since we will require that these effects are in balance, we now impose the condition that

$$\alpha = \epsilon^2. \tag{18}$$

At the next order, we obtain the system of equations

$$\zeta_T^{(1)} + hU_X^{(1)} = F^{(1)}, \tag{19}$$

$$U_T^{(1)} + g\zeta_X^{(1)} = G^{(1)}, \tag{20}$$

where

$$F^{(1)} = -\zeta_\tau^{(0)} - (U^{(0)}\zeta^{(0)})_X, \tag{21}$$

$$G^{(1)} = -U_\tau^{(0)} - U^{(0)}U_X^{(0)} + \frac{h^2}{3}U_{TXX}^{(0)}. \tag{22}$$

From (16) and (17), the inhomogeneous terms are functions of ξ and τ, and so, to leading order this system of equations reduces to

$$-c\zeta_\xi^{(1)} + hU_\xi^{(1)} = F^{(1)}, \tag{23}$$

$$-cU_\xi^{(1)} + g\zeta_\xi^{(1)} = G^{(1)}. \tag{24}$$

The homogeneous version of the system (23, 24) has a non-trivial solution, namely the right-propagating wave $\zeta^{(0)}, U^{(0)}$ given by (16). Hence the inhomogeneous system (23, 24) can have a solution only if the inhomogeneous terms on the left-hand side are orthogonal to the non-trivial solution of the homogeneous adjoint system. This is readily found to be (c, h) and so the required compatibility condition is

$$cF^{(1)} + hG^{(1)} = 0. \tag{25}$$

Next, we substitute the expressions in (16) into (25) and after some simplification get

$$A_\tau + \frac{3c}{2h}AA_\xi + \frac{ch^2}{6}A_{\xi\xi\xi} = 0. \tag{26}$$

Using the transformation (17), we recover the KdV equation (3) for the case of zero surface tension.

The same type of multiscale asymptotic expansion can be used to derive a KdV equation in many other physical systems. The key is the existence of a waveguide supporting a linear wave mode, whose dispersion relation for unidirectional sinusoidal waves, propagating along the waveguide (in the x-direction) with frequency ω and wavenumber k, has a long-wave expansion of the form.

$$\omega = ck - \delta k^3 + O(k^5). \tag{27}$$

A typical fluid variable, say $u(x, t, z)$, can be represented in the form

$$u = \alpha A(\xi, \tau)\phi(z) + O(\alpha^2, \alpha\epsilon^2), \tag{28}$$

Here the scaled variables ξ and τ are defined by (17), and $\phi(z)$ is a known structure function in the z-direction, where z is a coordinate across the waveguide. For instance, for water waves and when u is the horizontal velocity component, $\phi(z) = c/h$ (a constant) $(c = (gh)^{1/2}$ in this case), but in most physical systems, the dependence on z is not so simple, and is determined by an associated eigenvalue problem, which also determines the linear long-wave speed c. This is the situation for internal waves (see Grimshaw 2001 and Chapter 4). It is immediately clear that for linearized waves, the amplitude A will satisfy the linearization of the KdV equation (7). Thus the main task of the multiscale aysmptotic expansion is the determination of the nonlinear coefficient μ. This is accomplished by imposing the KdV balance condition (18) and constructing the equation for the second-order term in (28). This inevitably takes the form of a linear inhomogeneous system, whose homogeneous part is just the defining equation for the linear long-wave mode being considered. Hence, the inhomogeneous system requires a compatibility condition, which yields the required KdV equation (7).

2.2 Extensions

The KdV equation (7) forms the basis for many studies of solitary waves in fluids. It arises in the study of internal solitary waves (Chapter 4) and in the study of solitary waves in rotating fluids (Chapters 5 and 6). It is an integrable equation, the consequences of which are developed in Chapter 2. However, although it forms the basic paradigm, it is often necessary to modify it due to various practical concerns. For instance, strictly speaking it is valid only for weakly nonlinear waves, and there are circumstances when it is necessary to develop models capable of describing moderate and strongly nonlinear waves. This is often the case for internal solitary waves, and various higher-order KdV equations are described in Chapter 3. Here, it suffices to note that the general form for such an equation containing the next group of higher-order terms is

$$A_t + cA_x + \mu AA_x + \delta A_{xxx}$$
$$+ \{\mu_1 A^2 A_x + \delta_1 A_{xxxxx} + \nu_1 AA_{xxxx} + \nu_2 A_x A_{xx}\} = 0. \qquad (29)$$

The term in brackets is the set of higher-order terms, which are $O(\alpha)$ relative to the leading order KdV terms. In general, the form of these higher-order terms is not unique, as a change of variable through a near-identity transformation allows one, in general, to change any of the coefficients μ_1, δ_1, ν_1 or ν_2; in particular, it may be possible to asymptotically reduce (29) to the KdV equation. Thus, although (29) is usually not integrable, it is asymptotically integrable. These issues are developed in more detail in Chapter 4.

As we have already remarked, the linear terms of the KdV equation (7) (and also of the higher-order KdV equation (29)) replicate the long-wave expansion (27) of the linear dispersion relation. But this can be written in several different asymptotically equivalent forms: for instance, we could replace (27) with

$$\omega = \frac{ck}{1 + \delta k^2/c} + O(k^5). \qquad (30)$$

Translating this to the time domain, and reinstating the nonlinear term leads to the Benjamin–Bona–Mahoney (BBM) equation (Benjamin *et al.* 1972)

$$A_t + cA_x + \mu AA_x - \frac{\delta}{c} A_{xxt} = 0. \qquad (31)$$

Unlike the KdV equation (7), the BBM equation (31) is not integrable. Nevertheless, since it is as asymptotically equivalent to the KdV equation, and supports a solitary wave solution, it has sometimes been preferred to the KdV equation for numerical simulations because of its superior high wavenumber behaviour.

Both the KdV and the BBM equations are unidirectional. This is appropriate when studying the solitary wave solution and its interaction with other solitary waves propagating in the same direction with similar speeds (i.e. speeds close to the linear long-wave speed c). However, for oblique solitary wave interactions, one must either use the KP equation (6) when the interaction angle is small, or a Boussinesq equation otherwise. Boussinesq equations can take several forms; for surface waves, a popular form is (see, for instance, Mei 1983)

$$\zeta_t + \nabla \cdot \{(h + \zeta)\mathbf{U}\} = 0, \tag{32}$$

$$\mathbf{U}_t + \mathbf{U} \cdot \nabla \mathbf{U} + g\nabla\zeta - \frac{h^2}{3}\nabla\nabla \cdot \mathbf{U}_t = 0. \tag{33}$$

Even in the one-dimensional reduction, the equation system (32, 33) is not integrable. However, it can be written in a variety of asymptotically equivalent forms, and for some of these, the one-dimensional reduction is integrable (Kaup 1976; El *et al.* 2001). Like the KdV and BBM equations, the one-dimensional reduction supports a solitary wave solution. In general, the full two-dimensional version (32, 33) must be solved numerically. Similar Boussinesq equations have also been used to describe internal solitary waves (see Chapter 4).

The KdV equation (7) assumes that the solitary wave is propagating in a uniform background. However, there are situations when this is not the case. For instance, surface solitary waves may arise over a variable depth $h(x)$. In this situation, equation (3) is replaced by (see Johnson 1973)

$$\zeta_t + c\zeta_x + \frac{c_x}{2}\zeta + \frac{3c}{2h}\zeta\zeta_x + \frac{ch^2}{6}\delta\zeta_{xxx} = 0. \tag{34}$$

With the change of variable (compare (17))

$$\zeta = \epsilon^2 A, \quad \tau = \epsilon^3 s, \quad s = \int^x \frac{dx}{c(x)}, \quad \xi = \epsilon(s - t), \tag{35}$$

equation (34) transforms asymptotically to

$$A_\tau + \frac{c_\tau}{2c}A + \frac{3c}{2h}AA_\xi + \frac{ch^2}{6}A_{\xi\xi\xi} = 0. \tag{36}$$

Here, it has been assumed that c and h are slowly varying functions and depend on the slow variable τ. The extra term (compared to (3)) is to ensure the conservation of wave action flux,

$$\frac{d}{d\tau}\int_{-\infty}^{\infty} cA^2 \, d\xi = 0. \tag{37}$$

An immediate application of this is the demonstration that when a solitary wave propagates over a variable depth, its amplitude varies as h^{-1} (see Johnson 1973; Grimshaw 2001 and Chapter 2). When friction needs to be taken into account, which is usually the case for laboratory experiments, a dissipative term needs to be added to (3). For a turbulent bottom boundary layer, this takes the form

$$\zeta_t + c\zeta_x + \frac{3c}{2h}\zeta\zeta_x + \frac{ch^2}{6}\delta\zeta_{xxx} = -\frac{C_D c}{h^2}|\zeta|\zeta. \tag{38}$$

Here C_D is an empirical drag coefficient. From this equation it can be shown that the amplitude of a solitary wave decays with distance along the propagation path as $C_D x^{-1}$. Analogous variable-coefficient KdV equations have been developed to describe internal solitary waves in a variable background (Grimshaw 2001 and Chapter 4).

This monograph is about the solitary waves which occur in conservative systems, where there is no friction. This is the common and traditional scenario for solitary waves, which consequently can often be modelled by integrable model equations, such as the KdV equation described above and in Chapter 2. However, somewhat surprisingly, solitary waves can also occur in viscous fluid flows. A case which has been well studied is that of the flow of thin films under gravity. In that case, solitary waves can exist due to a balance between inertia, viscous forces and surface tension (Chang 1994; Balmforth 1995; Nepomnyashchy et al. 2002). In this, and in several other physical contexts (see, for instance, Nekorkin and Velarde 2002), a popular model equation is the so-called dispersion-modified Kuramoto–Sivashinsky equation,

$$A_t + \mu A A_x + \gamma_1 A_{xx} + \delta A_{xxx} + \gamma_2 A_{xxxx} = 0. \tag{39}$$

When $\delta = 0$, this reduces to the Kuramoto–Sivashinsky equation, where the second-order term with coefficient $\gamma_1 > 0$ is a low-wavenumber destabilizing term, and the fourth-order term with coefficient $\gamma_2 > 0$ is a high-wavenumber stabilizing term. From another point of view, equation (39) can be regarded as a KdV equation, perturbed with a low-wavenumber destabilizing mechanism and a high-wavenumber stabilizing mechanism. In essence, the solitary wave solutions of this equation exist due to a balance between these stabilizing and destabilizing effects. These solitary waves differ considerably from the classical KdV-type solitary waves. In particular, they are usually asymmetric and have exponential decay on one side, but oscillatory decay within an exponential envelope on the other side.

3 Solitary waves: bifurcation from the linear spectrum

Because solitary waves are required to decay in their tail regions, some information about their possible existence or otherwise can often be obtained by an examination of these tail regions, where, except in certain exceptional cases, a linearized analysis is applicable. One-dimensional steady solitary waves, propagating in the x-direction with speed c, are functions of $\xi = x - ct$, together with a set of other spatial transverse variables that define the modal structure. For instance, for surface or internal solitary waves, the dependence is on ξ and z (x is horizontal and z is vertical), and there is no dependence on the remaining horizontal variable y. In the tail region, where we assume that a linearized analysis holds, we seek solutions proportional to

$$\text{Re}\,\{\exp\,(ik(x - ct))\}. \tag{40}$$

The linearized equations will then yield the linear dispersion relation

$$c = c(k). \tag{41}$$

Whereas usually this dispersion relation (which may have several branches) is considered as an equation for c, given a real wavenumber k, for solitary wave tails, it needs to be considered as an equation for a complex-valued k given a real speed c. Indeed, it is immediately clear that if there exist real-valued solutions of (41) for the given value of c, then it is unlikely that the solitary wave can decay to its zero in its tail region. Instead, it is likely to be accompanied by a non-decaying co-propagating oscillatory wave field. This consideration leads to the notion that solitary waves generally can only exist in the gaps in the linear spectrum.

For instance, for water waves, the dispersion relation is (Lamb 1932; Mei 1983)

$$c^2 = \frac{g}{k}(1 + Bq^2)\tanh q, \quad q = kh, \tag{42}$$

where the Bond number $B = \Sigma/gh^2$ ($\rho\Sigma$ is the coefficient of surface tension and ρ is the water density, which has a value of 74 dynes/cm at 20°C). It may then be shown that solitary waves of the KdV type can exist only when either $B = 0$, $c^2 > gh$, or when $B > 1/3$, $c^2 < gh$, with a bifurcation from wavenumber zero ($k = 0$) and $c^2 = gh$ in both cases. Otherwise, when $0 < B < 1/3$, solitary waves can exist for $|c| < c_m$ where c_m^2 is the minimum value that c^2 can take in (42) as q takes all real values (in deep water, $|q| \to \infty$, $c_m^2 = 2(g\Sigma)^{1/2}$ and occurs at $|k| = k_m = (g/\Sigma)^{1/2}$). These solitary waves bifurcate at a finite wavenumber k_m and from the speed c_m, and have decaying oscillations in their tail regions. They are envelope solitary waves of a quite different kind from the aforementioned KdV-type solitary waves and will be discussed in detail in Chapter 7 (see also Dias and Bridges 2005 for a recent review).

This approach has recently been developed into the basis of a rigorous approach to finding solitary waves, often called the 'dynamical systems' method (see Grimshaw and Iooss 2003 for application to a model system, and Dias and Iooss 2003 for a comprehensive review of the application to water waves). The method was initiated by Iooss and Kirchgassner (1990) in a study of the existence of surface solitary waves in water in the presence of surface tension. Since the emphasis in this monograph is on solitary waves that occur in conservative systems, which is the common and traditional scenario for solitary waves, we shall suppose that the underlying physical system is Hamiltonian (i.e. energy conserving) and reversible (i.e. there is a symmetry under the transformation $\xi \to -\xi$). In this case, it can be shown the solutions k of the dispersion relation (41) for each real value of c have the property that $-k$ and k^* (complex conjugate) are also solutions. It follows that generically the solutions form a quartet $(k, k^*, -k, -k^*)$, with an associated four-dimensional subspace for the corresponding wave mode. For solitary waves, we require solutions with $\mathrm{Im}(k) > 0\ (< 0)$ according as $\xi \to \infty\ (\to -\infty)$, in order to ensure that the solution decays to zero in its tail region. In the general case when $\mathrm{Im}(k) \neq 0$, we see that there are generically two such roots available as $\xi \to \infty$ and, due to the reversible symmetry, two other roots available as $\xi \to -\infty$. Thus, for the corresponding wave mode, as $\xi \to \infty$, two boundary conditions are needed at each of $\pm\infty$. This count is consistent with the existence of a solitary wave solution, which from this dynamical systems point of view is a homoclinic orbit.

To make further progress, we now consider how this quartet structure may change as some system parameter is varied. Bifurcations arise when two solutions for k coalesce, for which the necessary condition is that $\partial c/\partial k = 0$ simultaneously with (41). When this occurs at a real value of k, it is equivalent to the condition that $c = c_g$, where $c_g = \partial\omega/\partial k$ is the group velocity and $\omega = ck$ is the frequency. Generically, there are four possibilities:

1. Two values of k coalesce at $k = 0$ and the remaining two values are such that $k = \pm i\gamma$, where $\gamma > 0$ is real-valued.
2. Two values of k coalesce at $k = 0$ and the remaining two values are such that $k = \pm\beta$, where $\beta > 0$ is real-valued.
3. Two values of k coalesce at $k = \beta$ and the remaining two values coalesce at $k = -\beta$, where $\beta > 0$ is real-valued.
4. Two values of k coalesce at $k = i\gamma$ and the remaining two values coalesce at $k = -i\gamma$, where $\gamma > 0$ is real-valued.

We will show that case 1 corresponds to a KdV-type solitary wave, and that case 3 corresponds to an envelope solitary wave. Case 2 corresponds to a so-called generalized solitary wave, which does not decay at infinity, but instead is accompanied there by small-amplitude co-propagating oscillations. Case 4 has only rarely been studied and corresponds to a transition from a KdV-type solitary wave to an envelope solitary wave (but see Dias *et al.* 1996; Dias and Iooss 2003).

The full system may now be projected onto the appropriate four-dimensional subspace, and the resulting bifurcation can then be analysed within the framework of this subspace. Of course, rigorous results require a delicate and sophisticated justification of this process. Here, we shall instead briefly describe the structure of the subspace, which we suppose is represented by the four-vector $\mathbf{W}(\xi)$. It satisfies an equation of the form

$$\mathbf{W}_\xi = L(\mathbf{W}; \epsilon) + N(\mathbf{W}). \tag{43}$$

Here, $L(\mathbf{W}; \epsilon)$ is a linear operator and $N(\mathbf{W})$ contains all the nonlinear terms. The bifurcation parameter is ϵ and is such that the spectrum of L at $\epsilon = 0$ reproduces one of the cases (1–4) described above. That is, the eigenvalues $\lambda = ik$ of the linear operator $L(\mathbf{W}; 0)$ are either (1) $(0, 0, -\gamma, \gamma)$, (2) $(0, 0, i\beta, -i\beta)$, (3) $(i\beta, i\beta, -i\beta, -i\beta)$ or (4) $(-\gamma, -\gamma, \gamma, \gamma)$.

Let us first consider case 1. At the bifurcation point ($\epsilon = 0$), the linearized system (43) has a double-zero eigenvalue, and generically there is a corresponding single eigenvector $\mathbf{V}_0$ and a single generalized eigenvector $\mathbf{V}_1$. Small-amplitude solutions are then sought in the form

$$\mathbf{W} = A(\xi)\mathbf{V}_0 + B(\xi)\mathbf{V}_1 + \mathbf{W}^{(2)}. \tag{44}$$

Here A and B are real variables of $O(\alpha)$, and the leading terms form a two-dimensional subspace (A, B), while $\mathbf{W}^{(2)}$ is a small error term of $O(\alpha^2, \alpha\epsilon)$, where $\epsilon, \alpha \ll 1$ are both small parameters. Note that the two remaining eigenvalues $\mp\gamma$ play no role at the leading order here, since they correspond to strong exponential decay at infinity, and their effects are included in the small error term $\mathbf{W}^{(2)}$. Projection onto the two-dimensional subspace and a normal form analysis then reveal that (A, B) satisfy the system

$$\begin{aligned} A_\xi &= B, \\ B_\xi &= \epsilon A + \mu A^2 + \cdots, \end{aligned} \tag{45}$$

where μ is a real-valued coefficient, specific to the system being considered, and the omitted terms are $O(\alpha\epsilon^2, \alpha^2\epsilon, \alpha^3)$. The coefficient ϵ yields the perturbed eigenvalues $\pm\epsilon^{1/2}$ for $\epsilon > 0$, and $\pm i|\epsilon|^{1/2}$ for $\epsilon < 0$; the former case yields the solitary wave solution. Comparison with the dispersion relation (41) leads to the identification of ϵ as

$$\epsilon = -\frac{c - c(0)}{c_{kk}(0)/2}. \tag{46}$$

It follows that for solitary waves, $c > (<) c(0)$ according as $c_{kk}(0) < (>) 0$, as expected. When the error terms in (45) are omitted, the resulting system can be recognized as the steady-state KdV equation and has the well-known 'sech^2' solution (compare (2)). It is then a delicate and intricate task to establish that this solitary wave solution persists when the error terms are restored.

Next, consider case 2. At the bifurcation point ($\epsilon = 0$) the linearized system (43) again has a double-zero eigenvalue, with a corresponding single eigenvector $\mathbf{V}_0$ and a single generalized eigenvector $\mathbf{V}_1$. However, account must now be taken of the other two eigenvalues $\pm i\beta$, with their associated eigenvectors $\mathbf{V}_2$ and $\mathbf{V}_2^*$, since they do not now lead to decaying solutions at infinity. Small-amplitude solutions are sought in the form

$$\mathbf{W} = A(\xi)\mathbf{V}_0 + B(\xi)\mathbf{V}_1 + C(\xi)\mathbf{V}_2 + C^*(\xi)\mathbf{V}_2^* + \mathbf{W}^{(2)}. \tag{47}$$

Here C is a complex-valued variable, and the leading terms form a four-dimensional subspace (A, B, C), while $\mathbf{W}^{(2)}$ is again a small error term. Projection onto this four-dimensional subspace and a normal form analysis reveal that (A, B, C) satisfy the system

$$\begin{aligned} A_\xi &= B, \\ B_\xi &= \epsilon A + \mu A^2 + \nu |C|^2 + \cdots, \\ C_\xi &= i\gamma(1 + \delta A)C + \cdots. \end{aligned} \tag{48}$$

Here, μ, ν and δ are real-valued coefficients specific to the system being considered, and the omitted terms are small error terms as above. When the error terms are omitted, the system is integrable. Indeed in that limit, $|C| = C_0$ is a constant, and after a change of origin, the system reduces to the same form as (45) in case 1. Thus, for the case $\epsilon > 0$ (when case 1 is a KdV-type solitary wave), the solution is a one-parameter family of circles of homoclinic-to-periodic solutions, with $|C| = C_0$ constant and $(A, B) \to (A_0,)$ as $\xi \to \pm\infty$, where A_0 is a real constant determined by solving $\epsilon A_0 + \mu A_0^2 + \nu C_0^2 = 0$. In less technical language, the solution is a generalized solitary wave, which typically has a 'sech^2' core and decays at infinity to non-zero oscillations of constant amplitude C_0 and wavenumber γ. A delicate analysis of the full system (43) with the small error terms shows that at least two of these solutions persist; the minimal amplitude C_0 being exponentially small, i.e. $O(\exp(-K/|\epsilon|^{1/2}))$, where K is a positive real constant (see, for instance, Lombardi 2000; Grimshaw and Iooss 2003). Although such waves are permissible as solutions of the steady-state equations, they have infinite energy and their associated group velocity is inevitably inward at one end and outward at the other end. Hence, they cannot be realized in a physical system from any localized initial condition. Instead, such initial conditions will typically generate a one-sided generalized solitary wave, whose central core is accompanied by small-amplitude outgoing waves on one side only. Such waves cannot be steady and instead will slowly decay with time (see, for instance, Akylas and Grimshaw 1992; Grimshaw et al. 1994; Grimshaw and Joshi 1995; Boyd 1998).

For water waves, for which the dispersion relation is (42), these two cases imply that pure solitary waves of elevation exist for $B = 0$ and of depression for $B > 1/3$, while generalized solitary waves arise whenever $0 < B < 1/3$ (see Amick and Kirchgassner 1989; Iooss and Kirchgassner 1990, 1992; Dias and Iooss 2003 for a recent review which includes references to much of the earlier literature on this topic). Chapter 3 contains a detailed account of such surface solitary waves. For the case of generalized solitary waves, there is always the possibility that the amplitude of the oscillations is zero, and the solution then reduces to a pure solitary, called an 'embedded' solitary wave. There are now many examples of such embedded solitary waves arising in various physical systems, notably for internal waves (see, for instance, Aklyas and Grimshaw 1992; Michallet and Dias 1999) but from numerical (Champneys et al. 2002) and analytical (Sun and Shen 1993; Sun 1999) studies, it would seem that they do not arise in the water wave context. This dynamical systems approach to finding solitary waves has also been applied to interfacial waves, where again the linear dispersion relation holds the key to where solitary waves can be found (see Dias and Iooss 2003 for details).

Finally we consider case 3. In this case, there is a double eigenvalue $\lambda = i\beta$ with generically a corresponding single eigenvector $\mathbf{V}_0$ and a single generalized eigenvector $\mathbf{V}_1$, while the complex conjugate double eigenvalue $\lambda = -i\beta$ has corresponding complex conjugate eigenvectors. Small-amplitude solutions are now sought in the form

$$\mathbf{W} = A(\xi)\mathbf{V}_0 + B(\xi)\mathbf{V}_1 + A^*(\xi)\mathbf{V}_0^* + B^*(\xi)\mathbf{V}_1^* + \mathbf{W}^{(2)}. \tag{49}$$

Here A and B are complex-valued variables, forming a four-dimensional subspace while $\mathbf{W}^{(2)}$ is again a small error term. Projection onto this subspace and a normal form analysis reveal that

$$A_\xi = i\beta A + B + iAP(\epsilon, |A|^2, K) + \cdots ,$$
$$B_\xi = i\beta B + i BP(\epsilon, |A|^2, K) + AQ(\epsilon, |A|^2, K) + \cdots . \tag{50}$$

where

$$K = i(AB^* - A^*B), \tag{51}$$

Here P and Q are real-valued polynomials of degree 1, i.e. we may write

$$P(\epsilon, |A|^2, K) = \epsilon + \nu_1 |A|^2 + \nu_2 K,$$
$$Q(\epsilon, |A|^2, K) = 2\epsilon\beta + \mu_1 |A|^2 + \mu_2 K, \tag{52}$$

where all coefficients are real-valued (see Grimshaw and Iooss 2003; Dias and Iooss 2003). The truncated system, with the error terms omitted, is integrable, as there are two integrals, i.e. K and H are both constants, where

$$H = |B|^2 - \left(2\epsilon\beta |A|^2 + \frac{\mu_1}{2}|A|^4 + \mu_2 K |A|^2\right). \tag{53}$$

For a solitary wave solution, we must have $K = H = 0$ and it then follows that

$$|A|_\xi^2 = 2\epsilon\beta |A|^2 + \frac{\mu_1}{2}|A|^4. \tag{54}$$

Thus solitary wave solutions exist provided that $\epsilon > 0$ and that the nonlinear coefficient $\mu_1 < 0$. The condition $\epsilon > 0$ implies that the perturbed eigenvalues, $\lambda \approx i\beta + (2\epsilon\beta)^{1/2}$, have split off the imaginary axis and so provide the conditions needed for exponential decay at infinity; the condition $\mu_1 < 0$ depends on the particular physical system being considered. For instance, it holds for capillary–gravity waves with $0 < B < 1/3$ (Dias and Iooss 1993). The solution of the truncated system is

$$A = a \exp(i[\beta + \epsilon]\xi)\operatorname{sech}(\gamma\xi), \tag{55}$$

where $\gamma = (2\epsilon\beta)^{1/2}$ and $|a|^2 = -4\epsilon\beta/\mu_1$. This solution describes an envelope solitary wave, with a carrier wavenumber $\beta + \epsilon$ and an envelope described by the 'sech' function. These kind of solitary waves are described in Chapter 7 from a slightly different point of view. Here we note that the solution (55) contains an arbitrary phase in that the argument of the complex amplitude a is not determined, meaning that the location of the crests of the carrier wave vis-á-vis the maximum of the envelope (here located at $\xi = 0$) is arbitrary. However, restoration of the error terms leads to the result that only two of these solutions persist (see Grimshaw and Iooss 2003; Dias and Iooss 2003), namely, those for which a carrier wave crest or trough is placed exactly at the envelope maximum $\xi = 0$, so that the resulting solitary wave is either one of elevation or depression. This result requires very delicate analysis, but could be anticipated by noting that these are the only two solutions which persist under the symmetry transformation $\xi \rightarrow -\xi$.

References

Ablowitz, M.J. & Segur, H., *Solitons and the Inverse Scattering Transform*, SIAM Studies in Applied Mathematics 4, SIAM: Philadelphia, 1981.

Akylas, T.R. & Grimshaw, R.H.J., Solitary internal waves with oscillatory tails. *J. Fluid Mech.*, **242**, pp. 279–298, 1992.

Amick, C. & Kirchgassner, K., A theory of solitary waves in the presence of surface tension. *Arch. Rat. Mech. Anal.*, **105**, pp. 1–49, 1989.

Balmforth, N.J., Solitary waves and homoclinic orbits. *Ann. Rev. Fluid Mech.*, **27**, pp. 335–373, 1995.

Benjamin, T.B., Bona, J.L. & Mahoney, J.J., Model equations for long waves in nonlinear dispersive systems. *Phil. Trans. Roy. Soc. A*, **272**, pp. 47–78, 1972.

Boussinesq, M.J., Theórie de l'intumescence liquid appellée onde solitaire ou de translation, se propageant dans un canal rectangulaire. *Comptes Rendus Acad. Sci., Paris*, **72**, pp. 755–759, 1871.

Boussinesq, M.J., Essai sur la theorie des eaux courantes, *Memoires presentees par diverse savants a l'Academie des Sciences Inst. France, Series 2*, **23**, pp. 1–680, 1877.

Boyd, J.P., Weakly nonlocal solitary waves and beyond-all-order asymptotics. *Mathematics and Its Applications*, vol. 442, Kluwer: Dordecht, 1998.

Champneys, A.R., Vanden-Broeck, J.-M. & Lord, G.J., Do true elevation gravity-capillary solitary waves exist? A numerical investigation. *J. Fluid Mech.*, **454**, pp. 403–417, 2002.

Chang, H.-C., Wave evolution on a falling film. *Ann. Rev. Fluid Mech.*, **26**, pp. 103–136, 1994.

Dias, F. & Bridges. T., Weakly nonlinear wave packets and the nonlinear Schrodinger equation (Chapter 2). *Nonlinear Waves in Fluids: Recent Advances and Modern Applications*, CISM Courses and Lectures No. 483, ed. R. Grimshaw, Springer: Wien and New York, pp. 29–67, 2005.

Dias, F. & Iooss. G., Capillary-gravity solitary waves with damped oscillations. *Physica D*, **65**, pp. 399–423, 1993.

Dias, F. & Iooss, G., Water waves as a spatial dynamical system. *Handbook of Mathematical Fluid Dynamics*, vol. 2, eds. S. Friedlander & D. Serre, Elsevier (North Holland), pp. 443–499, 2003.

Dias, F., Menasce, D. & Vanden-Broeck, J.-M., Numerical study of capillarygravity solitary waves. *Eur. J. Mech. B/Fluids*, **15**, pp. 17–36, 1996.

Dodd, R.K., Eilbeck, J.C., Gibbon, J.D. & Morris, H.C., *Solitons and nonlinear wave equations*, Academic Press: London, 1982.

Drazin, P.G. & Johnson, R.S., *Solitons: An Introduction*, Cambridge University Press: Cambridge, 1989.

El, G.A., Grimshaw, R.H.J. & Pavlov, M.V., Integrable shallow water equations and undular bores. *Stud. Appl. Math.*, **107**, pp. 157–186, 2001.

Gardner, C.S., Greene, J.M., Kruskal, M.D. & Miura, R.M., Method for solving the Korteweg-de Vries equation. *Physical Review Letters*, **19**, pp. 1095–1097, 1967.

Grimshaw, R., Internal solitary waves (Chapter 1). *Environmental Stratified Flows*, ed. R. Grimshaw, Kluwer: Boston, pp. 1–28, 2001.

Grimshaw, R., Korteweg-de Vries equation (Chapter 1). *Nonlinear Waves in Fluids: Recent Advances and Modern Applications*, CISM Courses and Lectures No. 483, ed. R. Grimshaw, Springer: Wien and New York, pp. 1–28, 2005.

Grimshaw, R. & Joshi, N., Weakly non-local solitary waves in a singularly perturbed Korteweg-de Vries equation. *SIAM J. Appl. Math.*, **55**, pp. 124–135, 1995.

Grimshaw, R. & Iooss, G., Solitary waves of a coupled Korteweg-de Vries system. *Mathematics and Computers in Simulation*, **62**, pp. 31–40, 2003.

Grimshaw, R., Malomed, B. & Benilov, E., Solitary waves with damped oscillatory tails: an analysis of the fifth-order Korteweg de-Vries equation. *Physica D*, **77**, pp. 473–485, 1994.

Iooss, G. & Kirchgassner, K., Bifurcation d'ondes solitaires en preśence d'une faible tension superficielle. *C.R. Acad. Sci, Paris*, **311**, pp. 265–268, 1990.

Iooss, G. & Kirchgassner, K., Water waves for small surface tension: an approach via normal form. *Proc. Roy. Soc. Edinburgh*, **122A**, pp. 267–299, 1992.

Johnson, R.S., On the development of a solitary wave over an uneven bottom. *Proc. Camb. Phil. Soc.*, **73**, pp. 183–203, 1973.

Kadomtsev, B.B. & Petviashvili, V.I., On the stability of solitary waves in weakly dispersing media. *Sov. Phys. Doklady*, **15**, pp. 539–541, 1970.

Kaup, D.J., A higher order wave equation and a method for solving it. *Progress Theor. Phys.*, **54**, pp. 396–408, 1976.

Korteweg, D.J. & de Vries, H., On the change of form of long waves advancing in a rectangular canal, and on a new type of long stationary waves. *Philosophical Magazine*, **39**, pp. 422–443, 1895.

Lamb, H., *Hydrodynamics*, Cambridge University Press, Cambridge, 1932.

Lombardi, E., *Oscillatory Integrals and Phenomena Beyond All Algebraic Orders: With Applications to Homoclinic Orbits in Reversible Systems*, Lecture Notes in Mathematics, Vol. 1741, Springer: New York, 2000.

Mei, C.C., *The Applied Dynamics of Ocean Surface Waves*, Wiley: New York, 1983.

Michallet, H. & Dias. F., Numerical study of generalized interfacial solitary waves. *Phys. Fluids*, **11**, pp. 1502–1511, 1999.

Miles, J.W., Solitary waves. *Annual Review of Fluid Mechanics*, **12**, pp. 11–43, 1980.

Nepomnyashchy, A.A., Velarde, M.G. & Colinet, P., *Interfacial Phenonema and Convection*, Chapman & Hall/CRC (Pittman Series # 124): London, 2002.

Nekorkin, V.I. & Velarde, M.G., *Synergetic Phenomena in Active Lattices: Patterns, Waves, Solitons, Chaos*, Springer-Verlag: Berlin, 2002.

Pego, R.L. & Weinstein, M.J., Convective linear stability of solitary waves for Boussinesq equations. *Studies Appl. Math.*, **99**, pp. 311–375, 1997.

Rayleigh, Lord, On waves. *Phil. Mag.* **1**, pp. 257–279, 1876.

Russell, J.S., Report on waves. *Proceedings of the 14th meeting of the British Association for the Advancement of Science*, John Murray: London, pp. 311–390, 1844.

Sun, S.M., Nonexistence of truly solitary waves in water with small surface tension. *Proc. R. Soc. Lond.*, **A455**, pp. 2191–2228, 1999.

Sun, S.M. & Shen, M.C., Exponentially small estimate for the amplitude of capillary ripples of a generalized solitary wave. *J. Math. Anal. Appl.*, **172**, pp. 533–566, 1993.

Zabusky, N.J. & Kruskal, M.D., Interactions of solitons in a collisionless plasma and the recurrence of initial states. *Physical Review Letters*, **15**, pp. 240–243, 1965.

CHAPTER 2

Korteweg–de Vries equation: solitons and undular bores

G.A. El

Department of Mathematical Sciences, Loughborough University, Loughborough, UK.

Abstract

The Korteweg–de Vries (KdV) equation is a fundamental mathematical model for the description of weakly nonlinear long wave propagation in dispersive media. It is known to possess a number of families of exact analytic solutions. Two of them – solitons and nonlinear periodic travelling waves – are of particular interest from the viewpoint of fluid dynamics applications as they occur as typical asymptotic outcomes in a broad class of initial/boundary-value problems. Two different major approaches have been developed in the last four decades to deal with the problems involving solitons and nonlinear periodic waves: inverse scattering transform and the Whitham method of slow modulations. We review these methods and show the relations between them. Emphasis is made on solving the KdV equation with large-scale initial data. In this case, the longtime evolution leads to the formation of an expanding undular bore, a modulated travelling wave connecting two different non-oscillating flows. Another problem considered is the propagation of a solitary wave through a variable environment in the framework of the variable-coefficient KdV equation. If the background environment varies slowly, the solitary wave deforms adiabatically and an extended small-amplitude trailing shelf is generated ensuring the conservation of mass. On a long time scale, the trailing shelf evolves, via an intermediate stage of an undular bore, into a secondary soliton train.

1 Introduction

The Korteweg–de Vries (KdV) equation

$$u_t + \alpha u u_x + \beta u_{xxx} = 0, \qquad (1)$$

is a universal mathematical model for the description of weakly nonlinear long wave propagation in dispersive media. Here $u(x, t)$ is an appropriate field variable and x and t are space coordinate and time, respectively. The coefficients α and β are determined by the medium properties and can be either constants or functions of x and t. An incomplete list of physical applications of the KdV equation includes shallow-water gravity waves, ion–acoustic waves in collisionless plasma, internal waves in the atmosphere and ocean, and waves in bubbly fluids. This broad range of applicability is explained by the fact that the KdV equation (1) describes a combined effect of the lowest-order, quadratic, nonlinearity (term uu_x) and the simplest long-wave dispersion (term u_{xxx}). One can find derivations of the KdV equation for different physical contexts in the books by Dodd *et al.* (1982), Drazin and Johnson (1989), Newell (1985), Karpman (1975) and many others.

Although equation (1) with constant coefficients was originally derived in the second half of the 19th century, its real significance as a fundamental mathematical model for the generation and propagation of long nonlinear waves of small amplitude has been understood only after the seminal works of Zabusky and Kruskal (1965), Gardner, Greene, Kruskal and Miura (1967) and Lax (1968). These authors showed that the KdV equation (unlike a 'general' nonlinear dispersive equation) can be solved *exactly* for a broad class of initial/boundary conditions and, importantly, the solutions often contain a combination of localized wave states, which preserve their 'identity' in the interactions with each other, pretty much as classical particles do. In the longtime asymptotic solutions, such localized states represent solitary waves, well known due to the original works of Russel, Boussinesq, Rayleigh, and Korteweg and de Vries. Such solitary wave solutions of the KdV equation have been called *solitons* by Zabusky and Kruskal (1965) owing to their unusual particle-like behaviour in the interactions with other solitary waves and nonlinear radiation. It turned out that the KdV equation possesses the family of N-soliton solutions where $N \in \mathbb{N}$ can be arbitrary. Later, similar multisoliton solutions have been found for other equations, which share with the KdV equation the remarkable property of *complete integrability*.

2 Periodic travelling wave solutions and solitary waves

We start with an account of the elementary properties of the travelling wave solutions of the KdV equation with constant coefficients α and β. Making in (1) a simple change of variables,

$$u' = \frac{1}{6}\alpha u, \quad x' = \frac{x}{\sqrt{\beta}}, \quad t' = \frac{t}{\sqrt{\beta}}, \tag{2}$$

we, on omitting primes, cast the KdV equation into its canonical dimensionless form

$$u_t + 6uu_x + u_{xxx} = 0. \tag{3}$$

We shall look for a solution of (3) in the form of a single-phase periodic wave of a permanent shape, i.e. in the form $u(x, t) = u(\theta)$, where $\theta = x - ct$ is the travelling phase and c = constant is the phase velocity. For such solutions, the KdV equation reduces to the ordinary differential equation

$$-cu_\theta + 6uu_\theta + u_{\theta\theta\theta} = 0, \tag{4}$$

which is integrated twice (the second integration requires an integrating factor u_θ) to give

$$\frac{1}{2}(u_\theta)^2 = (u - b_1)(u - b_2)(b_3 - u) \equiv G(u), \tag{5}$$

where $b_3 \geq b_2 \geq b_1$ are constants and

$$c = 2(b_1 + b_2 + b_3). \tag{6}$$

Equation (5) describes the periodic motion of a 'particle' with co-ordinate u and time θ in the potential $-G(u)$. Since $G(u) > 0$ for $u \in [b_2, b_3]$, the 'particle' oscillates between the endpoints b_2 and b_3 and the period of oscillations (the wavelength of the travelling wave $u(x - ct)$) is

$$L = \int_0^L d\theta = \sqrt{2} \int_{b_2}^{b_3} \frac{du}{\sqrt{G(u)}} = \frac{2\sqrt{2}K(m)}{(b_3 - b_1)^{1/2}}, \tag{7}$$

where $K(m)$ is the complete elliptic integral of the first kind and m is the modulus,

$$m = \frac{b_3 - b_2}{b_3 - b_1}, \quad 0 \leq m \leq 1. \tag{8}$$

The wavenumber and the frequency of the travelling wave $u(x, t)$ are

$$k = 2\pi/L, \quad \omega = kc. \tag{9}$$

Equation (5) is integrated in terms of the Jacobian elliptic cosine function $\text{cn}(\xi; m)$ (see, for instance, Abramowitz and Stegun 1965), to give

$$u(x, t) = b_2 + (b_3 - b_2)\text{cn}^2\left(\sqrt{2(b_3 - b_1)}(x - ct - x_0); m\right), \tag{10}$$

where x_0 is an initial phase. Solution (10) is often referred to as the *cnoidal wave*. According to the properties of elliptic functions, when $m \to 0$ ($b_2 \to b_3$), the cnoidal wave converts into the vanishing amplitude harmonic wave

$$u(x, t) \approx b_3 - a \sin^2[k_0(x - c_0 t - x_0)], \quad a = b_3 - b_2 \ll 1, \tag{11}$$

where $k_0 = k(b_1, b_3, b_3)$ and $c_0 = c(b_1, b_3, b_3)$. The relationship between c_0 and k_0 is obtained from equations (6) and (9) considered in the limit $b_2 \to b_3$:

$$c_0 = 6b_3 - k_0^2, \tag{12}$$

and agrees with the KdV linear dispersion relation $\omega(k) = 6ku_0 - k^3$ for linear waves propagating against the background $u_0 = b_3$.

When $m \to 1$ (i.e. $b_2 \to b_1$), the cnoidal wave (10) turns into a solitary wave

$$u_s(x, t) = b_1 + a_s \text{sech}^2 \left[\sqrt{a_s/2}(x - c_s t - x_0) \right], \tag{13}$$

whose speed of propagation $c_s = c(b_1, b_1, b_3)$ is connected with the amplitude $a_s = b_3 - b_1$ by the relation

$$c_s = 6b_1 + 2a_s. \tag{14}$$

Here, $u = b_1$ is a background flow: $u_s(x, t) \to b_1$ as $|x| \to \infty$. We note that, although the background flow in (11) and (13) can be eliminated by a passage to a moving reference frame and using the invariance of the KdV equation with respect to Galilean transformation,

$$u \to u + d, \quad x \to x - 6dt, \quad d = \text{constant}. \tag{15}$$

It is instructive to retain the full set of free parameters in the solution as they will be important in the study of the slowly varying solutions of the KdV equation, where b_1 and b_3 are no longer constants and, therefore, cannot be eliminated by transformation (15).

To sum up, the cnoidal waves form a *three-parameter* family of the KdV solutions, while the linear waves and solitary waves are characterised by only *two* independent parameters (with an account of background flow). We remark that the asymptotic solutions (11) and (13) could be obtained directly from the basic equation (5), which is easily integrated in terms of elementary functions when $b_2 \to b_3$ or $b_2 \to b_1$.

3 Inverse scattering transform method and solitons

Although the existence of the *particular* permanent shape travelling wave solutions such as (10) and (13) for a nonlinear partial differential equation is a non-trivial fact on its own, these solutions would have had very limited applicability if they would not appear in some reasonable class of initial-value or boundary-value problems. The real significance of these particular solutions becomes clear when one realises that solitary waves and periodic travelling waves naturally occur in the asymptotic solutions of a broad class of initial-value problems for the KdV equation and, moreover, methods exist enabling one to construct these solutions analytically. In this section, we will discuss some remarkable properties of the KdV solitary waves and the method for solving the problems involving their formation and evolution.

We will be interested in solving the KdV equation (3) in the class of functions decaying sufficiently fast together with their first derivatives far from the origin. With this aim in view, we consider the initial data

$$u(x, 0) = u_0(x), \quad u_0(x) \to 0, \quad u_0'(x) \to 0 \quad \text{as } |x| \to \infty. \tag{16}$$

The properties we will consider and the existence of the method of exact integration of the Cauchy problem (3), (16) reflect the fundamental fact of complete integrability of the KdV equation. Although the notion of complete integrability for a nonlinear partial differential equation has an exact mathematical meaning, which is formulated in terms of dynamics of infinite-dimensional Hamiltonian systems (see, for instance, Novikov *et al.* 1984; Newell 1985), we, for practical purposes of this text, will broadly mean by this the 'solvability' of a certain class of initial/boundary-value problems. Still, it is instructive to mention that the integrability of a finite-dimensional Hamiltonian system is intimately connected with the existence of 'higher than usual' number of conserved quantities.

3.1 Conservation laws

First, we note that the KdV equation (3) can be represented in the form of a *conservation law*

$$\frac{\partial u}{\partial t} + \frac{\partial}{\partial x}(3u^2 + u_{xx}) = 0. \tag{17}$$

Indeed, equation (17) implies conservation of the integral 'mass',

$$\frac{d}{dt} \int_{-\infty}^{+\infty} u dx = 0, \tag{18}$$

provided the function $u(x,t)$ vanishes together with its spatial derivatives as $x \to \pm\infty$ (another class of admissible functions is provided by periodic functions; in this case the integral in (18) should be taken over the period). Using simple algebra, one can also obtain conservation equations for 'momentum',

$$\frac{\partial}{\partial t} \frac{u^2}{2} + \frac{\partial}{\partial x}\left(2u^3 + uu_{xx} - \frac{1}{2}u_x^2\right) = 0, \tag{19}$$

and for 'energy',

$$\frac{\partial}{\partial t}\left(u^3 - \frac{1}{2}u_x^2\right) + \frac{\partial}{\partial x}\left(\frac{9}{2}u^4 + 3u^2u_{xx} + u_tu_x + \frac{1}{2}u_{xx}^2\right) = 0. \tag{20}$$

Now, we will show that the KdV equation actually possesses an *infinite number* of conservation laws. For that, we, following Gardner, introduce the differential substitution

$$u = w + i\epsilon w_x + \epsilon^2 w^2, \tag{21}$$

where ϵ is an arbitrary parameter. Setting (21) into the KdV equation (3) we obtain

$$u_t + 6uu_x + u_{xxx} = \left(1 + 2\epsilon^2 w + i\epsilon\frac{\partial}{\partial x}\right)(w_t + (3w^2 + 2\epsilon^2 w^3 + w_{xx})_x) = 0. \tag{22}$$

Now, it follows from (22) that if $w(x,t)$ satisfies the conservation equation

$$w_t + (3w^2 + 2\epsilon^2 w^3 + w_{xx})_x = 0, \tag{23}$$

then $u(x, t)$ is the solution of the KdV equation (3). We represent w in the form of an infinite asymptotic series in powers of ϵ (we do not require formal convergence though),

$$w \sim \sum_{n=0}^{\infty} \epsilon^n w_n. \tag{24}$$

Then, solving (21) by iterations for $\epsilon \ll 1$ we subsequently get:

$$w_0 = u, \quad w_1 = u_x, \quad w_2 = u^2 - u_{xx}, \quad \dots . \tag{25}$$

Now, setting (24) and (25) into the conservative equation (23), we obtain an infinite number of KdV conservation laws as coefficients at even powers of ϵ. It can be shown that the conservation laws corresponding to odd powers of ϵ represent the x-derivatives of the conservation laws corresponding to preceding even powers and thus, do not carry any additional information. The values w_{2n}, $n = 0, 1, 2, \dots$, are often called the *Kruskal integrals*.

Using the infinite set of the Kruskal integrals and the Hamiltonian structure, it was established that the KdV equation represents an infinite-dimensional integrable system. Practical realisation of this 'abstract' integrability is achieved through the inverse scattering transform method.

3.2 Lax pair

The method of integrating the KdV equation, discovered by Gardner, Greene, Kruskal and Miura (1967) and put into general mathematical context by Lax (1968), is based on the possibility of representing equation (3) as a compatibility (integrability) condition for two *linear* differential equations for the *same* auxiliary function $\phi(x, t; \lambda)$:

$$L\phi \equiv \left(-\partial_{xx}^2 - u\right)\phi = \lambda\phi, \tag{26}$$

$$\phi_t = A\phi \equiv \left(-4\partial_{xxx}^3 - 6u\partial_x - 3u_x + C\right)\phi \tag{27}$$

$$= (u_x + C)\phi + (4\lambda - 2u)\phi_x. \tag{28}$$

Here, λ is a complex parameter and $C(\lambda, t)$ is determined by the normalisation of ϕ. Equation (26) with appropriate boundary conditions constitutes the spectral problem and equation (27) the evolution problem. Direct calculation shows that the compatibility condition $(\phi_{xx})_t = (\phi_t)_{xx}$ yields the KdV equation (3) for $u(x, t)$ provided

$$\lambda_t = 0, \tag{29}$$

i.e. the evolution according to the KdV equation preserves the spectrum λ of the operator L in (26). This *isospectrality* property is very important for further analysis.

The operators L and A in (26) and (27), respectively, are often referred to as the *Lax pair*. It can be seen that the KdV equation (3) can be represented in an operator form $L_t = LA - AL \equiv [LA]$. This Lax representation provides a route for constructing further generalisations by appropriate choice of the operators L and A. For instance, it is clear that given the L-operator (26), the A-operator in the Lax pair is determined up to an operator commuting with L, which makes it possible to construct a *hierarchy* of equations associated with the same spectral problem but having different evolution properties. Such 'higher' KdV equations play an important role in constructing the nonlinear multiperiodic (multiphase) solutions of the original KdV equation (3) (see Novikov *et al.* 1984).

3.3 Direct scattering transform and evolution of spectral data

We consider the KdV equation in the class of functions sufficiently rapidly decaying as $|x| \to \infty$. To be more precise, we require boundedness of the integral (Faddeev's condition)

$$\int_{-\infty}^{+\infty} (1 + |x|)|u(x)|dx < \infty, \tag{30}$$

which ensures applicability of the scattering analysis in the sequel. Now, we turn to the first Lax equation (26), which represents the *time-independent Schrödinger equation*. This equation plays a central role in quantum mechanics (see, for instance, Landau and Lifshitz 1977) and describes the behaviour of the wave function $\phi(x; \lambda)$ of a particle moving through the potential $V(x) = -u(x)$. In this interpretation, $E = -\lambda$ is the energy of the particle. For a given potential $-u(x)$ the problem is to find the *spectrum* of the linear operator L, i.e. a set $\{\lambda\}$ of admissible values for λ, and to construct the corresponding functions $\phi(x; \lambda)$. Depending on the concrete form of the potential $-u(x)$, there are two different types of such solutions characterised by different types of the spectral set $\{\lambda\}$.

3.3.1 Continuous spectrum $\lambda > 0$: *scattering solutions*

We introduce $k^2 = \lambda$, $k \in \mathbb{R}$ and, assuming that $u \to 0$ as $x \to \pm\infty$ such that condition (30) is satisfied, fix an asymptotic behaviour of the function $\phi(x; k^2)$ far from origin:

$$\phi \sim \exp(-ikx) + R(k)\exp(ikx) \quad \text{as } x \to +\infty, \tag{31}$$

$$\phi \sim T(k)\exp(-ikx) \quad \text{as } x \to -\infty. \tag{32}$$

This asymptotic solution of equation (26) describes scattering *from the right* of the incident wave $\exp(-ikx)$ on the potential $-u(x)$. Then $R(k)$ represents a *reflection coefficient* and $T(k)$, a *transmission coefficient*. These are called *scattering data* and the problem of their determination for a given potential constitutes a direct scattering problem: $u \mapsto \{R(k), T(k)\}$. We note that owing to analytic properties of the solutions of the Schrödinger equation (see, for instance, Dodd *et al.* 1982), the functions $R(k)$ and $T(k)$ are not independent and the scattering data can be characterised by a single function $R(k)$. In particular, we mention the physically transparent [total probability] relationship $|T|^2 + |R|^2 = 1$ following from the constancy of the Wronskian for two independent scattering solutions with asymptotic behaviour (31) and (32).

Let the potential $-u(x,t)$ evolve according to the KdV equation (3). Then the corresponding evolution of the scattering data is found by substituting equations (31) and (32) into the second Lax equation (27) (note that for the continuous spectrum, the parameter λ can always be viewed as a constant). As a result, assuming $u(x), u'(x) \to 0$ as $x \to \pm\infty$, we obtain

$$C(\lambda, t) = 4ik^3, \quad \frac{dR}{dt} = 8ik^3 R, \quad \frac{dT}{dt} = 0. \tag{33}$$

Hence,

$$R(k,t) = R(k,0)\exp\left(8ik^3 t\right), \quad T(k,t) = T(k;0). \tag{34}$$

3.3.2 Discrete spectrum $\lambda = \lambda_n < 0$: *bound states*

If the potential $-u(x)$ is sufficiently negative near the origin of the x-axis, the scattering problem (26) implies the existence of a finite number of bound states $\phi = \phi_n(x; \lambda)$, $n = 1, \ldots, N$ corresponding to the discrete admissible values of the spectral parameter $\lambda = \lambda_n = -\eta_n^2$, $\eta_n \in \mathbb{R}$, $\eta_1 > \eta_2 > \cdots > \eta_N$. We require the following asymptotic behaviour as $x \to \pm\infty$, consistent with equation (26) for $u \to 0$:

$$\phi_n \sim \beta_n \exp\left(-\eta_n x\right) \quad \text{as } x \to +\infty, \tag{35}$$

$$\phi_n \sim \exp\left(\eta_n x\right) \quad \text{as } x \to -\infty. \tag{36}$$

Thus, for the case of a discrete spectrum, we have an analog of the scattering transform: $u \mapsto \{\eta_n, \beta_n\}$. Again, we shall be interested in the evolution of the spectral parameters when the potential $-u(x,t)$ evolves according to the KdV equation (3).

We first substitute equation (36) into (28) to obtain $C = C_n = 4\eta_n^3$ (cf. (33)). Then, setting (35) into (28) and using the isospectrality condition (29) we obtain

$$\frac{d\beta_n}{dt} = 4\eta_n^3 \beta_n \quad \text{so that } \beta_n(t) = \beta_n(0)\exp\left(4\eta_n^3 t\right). \tag{37}$$

We note that the bound state problem can be viewed as the analytic continuation of the scattering problem, defined on the real k-axis, to the upper half of the complex k-plane. Then the discrete points of the spectrum are found as *simple* poles $k = i\eta_n$ of the reflection coefficient $R(k)$ and $R \to 1$ as $|k| \to \infty$ (see details in Dodd *et al.* 1982 for instance).

Thus, for a general potential $-u(x)$, decaying as in (30), we can introduce the direct scattering transform by the mapping

$$u \mapsto S = \{(\eta_n, \beta_n); R(k), \ k \in \mathbb{R}, n = 1, \ldots, N\}. \tag{38}$$

Now, if the potential $u(x,t)$ evolves according to the KdV equation (3), then the scattering data S evolve according to simple equations

$$\eta_n = \text{constant}, \quad \beta_n(t) = \beta_n(0)\exp\left(4\eta_n^3 t\right), \quad R(k,t) = R(k,0)\exp\left(8ik^3 t\right). \tag{39}$$

The equations in (39) are often referred to as the Gardner–Greene–Kruskal–Miura equations.

3.4 Inverse scattering transform: Gelfand–Levitan–Marchenko equation

It was established in the 1950s that the potential $-u(x)$ of the Schrödinger equation can be completely reconstructed from the scattering data S. The corresponding mapping $S \mapsto u$ is called *inverse scattering transform* (IST) and is accomplished through the Gelfand–Levitan–Marchenko (GLM) linear integral equation. The derivation of this equation is beyond the scope of this text and can be found elsewhere (see, for instance, Whitham 1974; Drazin and Johnson 1989; Scott 2003). Here, we present only the resulting formulae and show some of their important consequences.

We define the function $F(x)$ as

$$F(x,t) = \sum_{n=1}^{N} \beta_n^2(t) \exp\left(-\eta_n x\right) + \frac{1}{2\pi} \int_{-\infty}^{+\infty} R(k,t) \exp\left(ikx\right) dk. \qquad (40)$$

Then the potential $-u(x,t)$ is restored from the formula

$$u(x,t) = 2 \frac{\partial}{\partial x} K(x,x,t), \qquad (41)$$

where the function $K(x,y,t)$ is found from the linear integral (GLM) equation

$$K(x,y) + F(x+y) + \int_{x}^{+\infty} K(x,z)F(y+z)dz = 0 \qquad (42)$$

defined for any moment t.

Thus, we have the scheme of integration of the KdV equation by the IST method:

$$u(x,0) \mapsto S(0) \to S(t) \mapsto u(x,t). \qquad (43)$$

It is essential that at each step of this algorithm one has to solve a linear problem. One can notice that the described method of integration of the KdV equation is in many respects analogous to the Fourier method for integrating linear partial differential equations with the role of direct and inverse Fourier transform played by the direct and inverse scattering transform. Moreover, it can be shown (see, for instance, Ablowitz and Segur 1981) that for linear problems, the scattering transform indeed converts into the usual Fourier transform.

3.5 Reflectionless potentials and N-soliton solutions

There exists a remarkable class of potentials characterised by the zero reflection coefficient, $R(k) = 0$. Such potentials are called *reflectionless* and can be expressed in terms of elementary functions. We start with the simplest case $N = 1$. In this case, since $R(k) = 0$, we have from (40) and (39): $F(x,t) = \beta(0)^2 \exp\left(-\eta x + 8\eta^3 t\right)$, where $\beta \equiv \beta_1$ and $\eta \equiv \eta_1$. Then the solution of equation (42) can be sought in the form $K(x, y, t) = M(x, t) \exp(-\eta y)$. After simple algebra, we get

$$M(x, t) = \frac{-2\eta\beta(0)^2 \exp(-\eta x + 8\eta^2 t)}{2\eta + \beta(0)^2 \exp(-2\eta x + 8\eta^2 t)}. \tag{44}$$

As a result, we obtain from (41)

$$u = 2\eta^2 \text{sech}^2(\eta(x - 4\eta^2 t - x_0)), \tag{45}$$

which is just the solitary wave (13) of the amplitude $a_s = 2\eta^2$, propagating on a zero background ($b_1 = 0$) to the right with the velocity $c_s = 4\eta^2$ and having the initial phase

$$x_0 = \frac{1}{2\eta} \ln \frac{\beta(0)^2}{2\eta}. \tag{46}$$

For arbitrary $N \in \mathbb{N}$ and $R(k) \equiv 0$, we have from (40)

$$F(x) = \sum_{n=1}^{N} \beta_n(t)^2 \exp(-\eta_n x),$$

and, therefore, seek the solution of the GLM equation (42) in the form

$$K(x, y, t) = \sum_{n=1}^{N} M_n(x, t) \exp(-\eta_n y). \tag{47}$$

Now, on using (41), one arrives, after some algebra (see, for instance, Novikov et al. 1984), at the general representation for the reflectionless potential $V(x, t) = -u_N(x, t)$,

$$u_N(x, t) = 2 \frac{\partial^2}{\partial x^2} \ln \det A(x, t). \tag{48}$$

Here, A is the $N \times N$ matrix given by

$$A_{mn} = \delta_{mn} + \frac{\beta_n(t)^2}{\eta_n + \eta_m} e^{-(\eta_n + \eta_m)x}, \tag{49}$$

δ_{km} being the Kronecker delta.

The analysis of formulae (48) and (49) (see Karpman 1975; Novikov et al. 1984; Drazin and Johnson 1989) shows that for $t \to \pm\infty$, the solution of the KdV equation corresponding to the reflectionless potential can be asymptotically represented as a superposition of N single-soliton solutions propagating to the right and ordered in space by their speeds (amplitudes):

$$u_N(x, t) \sim \sum_{n=1}^{N} 2\eta_n^2 \text{sech}^2[\eta_n(x - 4\eta_n^2 t \mp x_n)], \quad \text{as } t \to \pm\infty, \tag{50}$$

where the amplitudes of individual solitons are given by $a_n = 2\eta_n^2$ and the positions $\mp x_n$ of the nth soliton as $t \to \mp\infty$ are given by the relationship (cf. (46) for a single soliton)

$$x_n = \frac{1}{2\eta_n} \ln \frac{\beta_n^2(0)}{2\eta_n} + \frac{1}{2\eta_n} \left\{ \sum_{m=1}^{n-1} \ln \left| \frac{\eta_n - \eta_m}{\eta_n + \eta_m} \right| - \sum_{m=n+1}^{N} \ln \left| \frac{\eta_n - \eta_m}{\eta_n + \eta_m} \right| \right\}. \quad (51)$$

One can infer from equation (50) that at $t \gg 1$, the tallest soliton with $n = N$ is at the front followed by the progressively shorter solitons behind, forming thus the triangle amplitude (velocity) distribution characteristic for non-interacting particles (see Whitham 1974). At $t \to -\infty$, we get the reverse picture. The full solution in (48) and (49) thus describes the interaction (collision) of N solitons at finite times. For this reason it is called N-soliton solution. The N-soliton solution is characterised by $2N$ parameters $\eta_1, \ldots, \eta_N$, $\beta_1(0), \ldots, \beta_N(0)$. Owing to isospectrality ($\eta_n =$ constant), the solitons preserve their amplitudes (and velocities) in the interactions; the only change they undergo is an additional phase shift $\delta_n = 2x_n$ due to collisions.

Say, for a two-soliton collision with $\eta_1 > \eta_2$ the phase shifts as $t \to +\infty$ are

$$\delta_1 = 2x_1 = \frac{1}{\eta_1} \ln \left| \frac{\eta_1 - \eta_2}{\eta_1 + \eta_2} \right|, \quad \delta_2 = -2x_2 = -\frac{1}{\eta_2} \ln \left| \frac{\eta_1 - \eta_2}{\eta_1 + \eta_2} \right|. \quad (52)$$

It follows from the formula (52) that, as a result of the collision, the taller soliton gets an additional shift forward by the distance $2x_1$, while the shorter soliton is shifted backwards by the distance $2x_2$. One should also keep in mind that the general formula (51) and its two-soliton reduction (52) are relevant only for sufficiently large times when individual solitons are sufficiently separated for the asymptotic representation (50) to be applicable.

In conclusion of this section, we note that one of the remarkable consequences of the formula (51) for the phase shifts is that the solitons in the N-soliton solution of the KdV equation interact only pairwise, i.e. the 'multi-particle' effects in the soliton interactions are absent.

3.6 Purely reflective potentials: nonlinear radiation

As opposed to the reflectionless potentials, the *purely reflective* potentials are characterised by zero transmission coefficient $T(k) \equiv 0$. This is evidently the case for all positive potentials $V(x) = -u_0(x) \geq 0$ characterised by a pure continuous spectrum. Now, one has to deal with the second term alone in formula (40). In this case, the general expression for the solution similar to the N-soliton solution is not available. An asymptotic analysis (see Dodd *et al.* 1982; Ablowitz and Segur 1981 and references therein) shows that, under the longtime evolution, the purely reflective potential transforms into the linear dispersive wave (the radiation) described locally by the linearised KdV equation but, unlike that in the solution of the initial-value problem for the linearised KdV equation, its amplitude decays at each fixed point at $x < 0$ with the rate greater or equal to $t^{-1/2}$ rather than $t^{-1/3}$. The detailed structure of this wave and its dependence on the initial data are very complicated. However, the local qualitative behaviour is physically transparent: the linear radiation propagates to the left with the velocity close to the group velocity $c_g = -3k^2$ of a linear wave packet, and the lowest rate of the amplitude decay is consistent with the momentum conservation law in the linear modulation theory (see Whitham 1974).

3.7 Evolution of an arbitrary decaying potential

Now, we are able to qualitatively describe an asymptotic evolution of an arbitrary decaying potential satisfying the condition (30). The spectrum of such a potential generally contains both discrete and continuous components. The discrete component is responsible for the appearance in the asymptotic solution of the chain of solitons ordered by the amplitudes and moving to the right. At the same time, the continuous component contributes to the linear dispersive wave propagating to the left. A simple *sufficient* condition for the appearance of at least one bound state in the spectrum (i.e. of a soliton in the solution) is

$$\int_{-\infty}^{+\infty} u_0(x)dx > 0. \tag{53}$$

Thus, the longtime asymptotic outcome of the general KdV initial-value problem for decaying initial data can be represented in the form

$$u(x,t) \sim \sum_{n=1}^{N} 2\eta_n^2 \text{sech}^2(\eta_n(x - 4\eta_n^2 t - x_n)) + \text{linear radiation}, \tag{54}$$

where the soliton amplitudes $a_n = 2\eta_n^2$ and the initial phases x_n, as well as the parameters of the radiation component, are determined from the scattering data for the initial potential. The upper bound for the number N of solitons in the solution can be estimated by the formula

$$N \leq 1 + \int_{-\infty}^{+\infty} |x||u_0(x)|dx. \tag{55}$$

Unfortunately, even the direct scattering problem can be solved explicitly only for very few potential forms. In most cases, one has to use numerical solutions or asymptotic estimates.

3.8 Semi-classical asymptotics in the IST method

One of the important cases where some explicit analytic results of rather general form become available occurs when the initial potential is a 'large-scale' function. Then, for positive $u_0(x)$, the Schrödinger operator (26) has a large number of bound states located close to each other so that the discrete spectrum can be characterised by a single continuous distribution function. In this case, an effective asymptotic description of the spectrum can be obtained with the use of the semi-classical Wentzel–Kramers–Brillouin (WKB) method (see, for instance, Landau and Lifshitz 1977).

We consider the KdV equation (3) with the large-scale positive initial data

$$u(x, 0) = u_0(x/L) > 0, \quad L \gg 1. \tag{56}$$

For simplicity, we assume that the initial function (56) has a form of a single positive bump satisfying an additional condition

$$\int_{-\infty}^{\infty} u_0^{1/2} dx \gg 1, \tag{57}$$

whose meaning will be clarified soon. An estimate following from equation (57) is $A^{1/2}L \gg 1$, where $A = \max(u_0)$. Assuming $A = \mathcal{O}(1)$, we introduce 'slow' variables $X = \epsilon x$ and $T = \epsilon t$, where $\epsilon = 1/L \ll 1$ is a small parameter, into equation (3) to get the small-dispersion KdV equation:

$$u_T + 6uu_X + \epsilon^2 u_{XXX} = 0, \quad \epsilon \ll 1 \tag{58}$$

with the initial condition

$$u(X, 0) = u_0(X) \geq 0, \tag{59}$$

where

$$u_0(X) \text{ is } C^1 \quad \text{and} \quad \int_{-\infty}^{\infty} u_0^{1/2} dX = \mathcal{O}(1). \tag{60}$$

The associated Schrödinger equation (26) in the Lax pair assumes the form

$$-\epsilon^2 \phi_{XX} - u\phi = \lambda\phi. \tag{61}$$

We note that in quantum mechanics, the role of ϵ in equation (61) is played by the Planck constant $\hbar$. According to the IST ideology, in order to construct the solution of the KdV equation (58) in the asymptotic limit $\epsilon \to 0$, we need to study the corresponding asymptotic behaviour of the scattering data for the initial potential $-u_0(X)$. The scattering data set S consists of two groups of parameters (see Section 3.3): eigenvalues $-\eta_n^2$ and norming coefficients β_n, $n = 1, \ldots, N$ characterising a discrete spectrum and the reflection coefficient $R(k)$ characterising a continuous spectrum.

The WKB analysis of the Schrödinger equation (61) yields that, for the potential $-u_0(X) \leq 0$ satisfying condition (57), the reflection coefficient is asymptotically zero,

$$\lim_{\epsilon \to 0} R(k) = 0, \tag{62}$$

while the eigenvalues $\lambda_n = -\eta_n^2$ ($n = 1, \ldots, N$ and $\eta_1 > \eta_2 > \cdots > \eta_N \geq 0$) are distributed in the potential range $-A \leq -\eta_n^2 \leq 0$ and the density of the distribution of η_ns is given by the formula (Weyl's law)

$$\phi(\eta) = \frac{1}{\pi\epsilon} \int_{X^-(\eta)}^{X^+(\eta)} \frac{\eta}{\sqrt{u_0(X) - \eta^2}} dX, \tag{63}$$

so that $\phi(\eta)d\eta$ is the number of η_ks in the interval $(\eta; \eta + d\eta)$. Here, the limits of integration $X^-(\eta) < X^+(\eta)$ are the turning points defined by $u_0(X^{\pm}) = \eta^2$. Formula (63) follows from the famous Bohr–Sommerfeld semi-classical quantisation rule (see Landau and Lifshitz 1977), which we represent in the form

$$\oint \sqrt{u_0(X) - \eta^2}\, dX = 2\pi\epsilon \left(n - \frac{1}{2}\right), \quad n = 1, 2, \ldots N, \tag{64}$$

so that the number of bound states in the spectral interval $(-A, -\eta^2)$, where $\lambda_n \leq -\eta^2 \leq \lambda_{n+1}$ and $\lambda_{N+1} \equiv 0$, is equal to n. Then, considering n as a continuous function of η, we have for the density of η_ks: $\phi(\eta) = |dn/d\eta|$, which yields (63). The integration in (64) is performed over the full period of the motion of the classical particle with the energy $-\eta^2$ in the potential well $-u_0(X)$. The total number of bound states N can be estimated by setting $\eta = 0$ in (64):

$$N \sim \frac{1}{\pi\epsilon} \int_{-\infty}^{+\infty} u_0^{1/2}(X)dX \gg 1. \tag{65}$$

The inequality in (65) is equivalent to the condition (57) and clarifies its physical meaning. The norming constants $\beta_n(0)$ of the scattering data in the semi-classical limit are given by the formulae (see Lax, Levermore and Venakides 1994)

$$\beta_n = \beta(\eta_n), \quad \beta(\eta) = \exp\{\chi(\eta)/\epsilon\}, \tag{66}$$

where

$$\chi(\eta) = \eta X^+(\eta) + \int_{X^+(\eta)}^{\infty} \left(\eta - \sqrt{\eta^2 - u_0(X)}\right) dX. \tag{67}$$

Now, we interpret the semi-classical scattering data (62)–(67) in terms of the solution $u(X, T)$ of the small-dispersion KdV equation (58). First of all, the relation (62) implies that the potential $-u_0(X)$ is *asymptotically reflectionless* and, hence, the initial data $u_0(X)$ can be approximated by the N-soliton solution (48), (49),

$$u_0(X) \approx u_N(X/\epsilon) \quad \text{for } \epsilon \ll 1, \tag{68}$$

where $N[u_0] \sim \epsilon^{-1}$ is given by (65) and the discrete spectrum is defined by (63), (66) and (67). Now, one can use the known N-soliton dynamics (see Section 3.5) for the description of the evolution of an arbitrary initial potential satisfying the condition (57). This observation served as a starting point in the series of papers by Lax , Levermore and Venakides (see their 1994 review and references therein), where the singular zero-dispersion limit of the KdV equation has been introduced and thoroughly studied. While the description of multisolitons at finite T turns out to be quite complicated in the zero-dispersion limit, the asymptotic behaviour as $T \to \infty$ can be easily predicted using formula (50), which implies that as $T \to \infty$ the outcome of the evolution will be a 'soliton train' consisting of N free solitons ordered by their amplitudes $a_n = 2\eta_n^2$, $n = 1, \ldots, N$ and propagating on a zero background. The number of solitons in the train having the amplitude within the interval $(a, a + da)$ is $f(a)da$, where the soliton amplitude distribution function $f(a)$ follows from Weyl's law (63):

$$ f(a) = \frac{1}{8\pi\epsilon} \oint \frac{dX}{\sqrt{u_0(X) - a/2}}. \tag{69} $$

The formula (69) was obtained for the first time by Karpman (1967). It follows from the Karpman formula that the range of soliton amplitudes in the train is

$$ 0 < a < 2A, \tag{70} $$

which means that the tallest soliton has the amplitude twice as big as the amplitude of the initial perturbation, $a_{max} = 2A$. The Karpman formula also allows one to determine the spatial distribution of solitons in the soliton train resulting from the initial perturbation $u_0(X)$. Indeed, as the speed of the soliton with the amplitude a moving on a zero background is $c_s = 2a$ (see (14)), its asymptotic position for $T \gg 1$ is $X \cong 2aT$, which implies that the solitons in the soliton train are spatially distributed according to a 'triangle' law

$$ a \cong X/2T, \qquad 0 < X/2T < 2A, \quad X, T \gg 1. \tag{71} $$

The number of waves in the interval $(X, X + dX)$ is determined from the balance relationship

$$ \kappa\, dX = f(a)da, \tag{72} $$

where $\kappa(X, T)$ is the spatial density of solitons (the soliton train wavenumber divided by 2π). Then, using (71) we obtain

$$ \kappa(X, T) \cong \frac{1}{2T}\, f\left(\frac{X}{2T}\right). \tag{73} $$

Whereas the longtime multisoliton dynamics in the semi-classical limit is simple enough, the corresponding behaviour at finite times T is quite non-trivial and reveals some remarkable features. As the studies of Lax, Levermore and Venakides 1994 showed, there is a certain critical time $T_b[u_0(X)]$, after which the multisoliton solution of the small-dispersion KdV equation (58), resulting from the bump-like initial data, asymptotically to the first order in ϵ manifests itself as a *cnoidal wave* (10) with the period and the wavelength scaled as ϵ and with the parameters b_j depending on the slow variables X and T. Moreover, Lax and Levermore obtained the evolution equations for the moments $\bar{u}(X, T)$, $\overline{u^2}(X, T)$ and $\overline{u^3}(X, T)$, which turned out to coincide with the *modulation equations* derived much earlier by Whitham (1965).

4 Whitham modulation equations

4.1 Whitham method

In the 1960s, G. Whitham developed an asymptotic theory to treat the problems involving periodic travelling wave solutions (cnoidal waves in the KdV equation context) rather than individual solitons. It is clear that the cnoidal wave solution (10) as such, similar to the plane monochromatic wave in linear wave theory, does not transfer any 'information' and does not solve any reasonable class of initial-value problems (except for the problem with the initial data in the form of a cnoidal wave). However, one can try to construct a *modulated cnoidal wave*, a nonlinear analog of a linear wave packet, which can presumably be an asymptotic outcome in some class of the nonlinear dispersive initial-value problems.

It is convenient to represent the periodic travelling wave solution (10) in an equivalent general form

$$u(x,t) = U_0(\tau; \mathbf{b}), \qquad \tau = kx - \omega t - \tau_0, \quad \mathbf{b} = (b_1, b_2, b_3), \qquad (74)$$

where the functions $k(\mathbf{b})$ and $\omega(\mathbf{b})$ are determined from (9), (6) and (7), and the function $U_0(\tau)$ is defined by the ordinary differential equation (ODE) $k^2(U_0')^2 = 2G(U_0)$ (an equivalent of (5)) and is 2π-periodic, $U_0(\tau + 2\pi; \mathbf{b}) = U_0(\tau; \mathbf{b})$; $\tau_0(\mathrm{mod}\, 2\pi)$ being an arbitrary initial angular phase. We introduce a *slowly mod-ulated* cnoidal wave by letting the constants of integration b_j be functions of x and t on a large spatio-temporal scale, i.e. $b_j = b_j(X, T)$, where $X = \epsilon x$, $T = \epsilon t$, and $\epsilon \ll 1$ is a small parameter. Now equation (10) (or (74)) no longer is an exact solution of the KdV equation (3). One can, however, require that $U_0(\tau, \mathbf{b}(X, T))$ satisfies the KdV equation *approximately*, i.e. to first order in ϵ. This requirement leads to a set of restrictions for the slowly varying functions $b_j(X, T)$, which are called *modulation equations*.

The modulation equations can be obtained by using an extension of the well-known multiple-scale perturbation method (see, for instance, Nayfeh 1981) to nonlinear partial differential equations (see Luke 1966; Grimshaw 1979; Dubrovin and Novikov 1989 and references therein).

We shall seek an asymptotic solution of the KdV equation in the form

$$u = u^{(0)} + \epsilon u^{(1)} + \epsilon^2 u^{(2)} + \cdots, \qquad (75)$$

where the leading term $u^{(0)}$ has the form (74) but with slowly varying parameters b_1, b_2 and b_3. To this end, we introduce an auxiliary phase function $S(X, T)$ and represent the terms $u^{(n)}$ of the decomposition (75) in the form

$$u^{(n)} = U_n(S(X, T)/\epsilon; \mathbf{b}(X, T)), \qquad (76)$$

where $U_n(S(X, T))$ are 2π-periodic functions, which depend smoothly on X and T. Then, for the leading term in (75) to have the form of the travelling wave (74), i.e. $u^{(0)} \to U_0(kx - \omega t; \mathbf{b})$ as $\epsilon \to 0$, one should require

$$S_X = k(\mathbf{b}(X, T)), \quad S_T = -\omega(\mathbf{b}(X, T)) \qquad (77)$$

(this is readily established by substituting (74) and (75)–(77) into the KdV equation (3) and comparing the coefficients of the resulting ODEs appearing to the leading order in ϵ). The compatibility condition $S_{XT} = S_{TX}$ yields the so-called wave conservation law

$$k_T + \omega_X = 0, \tag{78}$$

which is one of the modulation equations. The remaining two are obtained by considering the next leading order equation in the asymptotic chain occurring after the substitution of the expansion (75) into the KdV equation (3) with the account of the form of the leading periodic term U_0 (74). As a result, collecting the terms $\mathcal{O}(\epsilon)$, one arrives at the ODE

$$-\omega(U_1)_\tau + 6k(U_0 U_1)_\tau + k^3(U_1)_{\tau\tau\tau} = -\frac{\partial}{\partial T}U_0 - 6U_0\frac{\partial U_0}{\partial X}. \tag{79}$$

Since the right-hand side of equation (79) is a 2π-periodic function in τ, an unbounded growth of the solutions is expected due to resonances with the eigenfunctions of the linear operator on the left-hand side. To eliminate this unbounded growth, one should impose the orthogonality conditions,

$$\int_0^{2\pi} y_\alpha R d\tau = 0, \quad \alpha = 1, 2, \tag{80}$$

where $R(U_0, \partial_T U_0, \partial_X U_0)$ is the right-hand side of equation (79), and y_α are the periodic eigenmodes of the operator adjoint to the homogeneous operator on the left-hand side of equation (79). The adjoint equation has the form:

$$-\omega y_\tau + 6kU_0 y_\tau + k^3 y_{\tau\tau\tau} = 0. \tag{81}$$

One can see that there are indeed just two periodic solutions of equation (81): $y_1 = 1$ and $y_2 = U_0$. Equations (78) and (80) represent then the full set of the modulation equations for the three parameters $b_j(X, T)$.

There is a convenient alternative to the direct perturbation procedure outlined above. This alternative (but, of course, equivalent at the end) method was proposed by Whitham in 1965. The Whitham method of obtaining the modulation equations prescribes *averaging* any three of the KdV conservation laws $\partial_t P_j + \partial_x Q_j = 0$ (equations (18)–(20), for instance) over the period L of the travelling wave (10) (or over 2π-interval if one uses the solution in the form (74)). The averaging is made according to (5) as

$$\overline{F}(b_1, b_2, b_3) = \frac{1}{L}\int_0^L F(u(\theta; b_1, b_2, b_3))d\theta = \frac{k}{\pi}\int_{b_2}^{b_3}\frac{F(u)}{\sqrt{-G(u)}}du. \tag{82}$$

In particular, the mean value is calculated as

$$\bar{u} = b_1 + 2(b_3 - b_1)E(m)/K(m), \tag{83}$$

where $E(m)$ is the complete elliptic integral of the second kind. We now express the partial t- and x-derivatives as asymptotic expansions

$$\frac{\partial F}{\partial t} = \frac{dF}{d\theta} + \epsilon \frac{\partial F}{\partial T} + \mathcal{O}(\epsilon)^2, \quad \frac{\partial F}{\partial x} = \frac{dF}{d\theta} + \epsilon \frac{\partial F}{\partial X} + \mathcal{O}(\epsilon)^2. \tag{84}$$

Then, in view of periodicity of F in θ, the averages of the derivatives (84) are calculated as

$$\overline{\frac{\partial}{\partial t}F} = \epsilon \frac{\partial}{\partial T}\overline{F} + \mathcal{O}(\epsilon)^2, \quad \overline{\frac{\partial}{\partial t}F} = \epsilon \frac{\partial}{\partial T}\overline{F} + \mathcal{O}(\epsilon)^2. \tag{85}$$

Now, applying the averaging (82) to the conservation laws (18), (19) and (20) and passing to the limit as $\epsilon \to 0$, we arrive at the KdV modulation system in a conservative form

$$\frac{\partial}{\partial T}\overline{P}_j(b_1, b_2, b_3) + \frac{\partial}{\partial X}\overline{Q}_j(b_1, b_2, b_3) = 0, \quad j = 1, 2, 3, \tag{86}$$

where $\overline{P}_j(\mathbf{b})$ and $\overline{Q}_j(\mathbf{b})$ are expressed in terms of the complete elliptic integrals of the first and the second kind.

The equivalence of Whitham's method of averaging to the formal multiple-scale perturbation procedure is now rigorously established for a broad class of integrable equations including, of course, the KdV equation (see Dubrovin and Novikov 1989 and references therein). We also mention two non-trivial but easy to understand facts, which play an important role in the modulation theory:

1. all modulation systems obtained by averaging *any three* independent conservation laws from the infinite set available for the KdV equation are equivalent to the basic system (86);
2. the wave conservation equation (78) which was obtained here using multiple-scale expansions and which does not correspond to any particular averaged KdV conservation law, is consistent with the system of three averaged conservation laws (86) and, thus, can be used instead of any of them (see Whitham 1965).

It was discovered Whitham (1965) that on introducing symmetric combinations

$$r_1 = \frac{b_1 + b_2}{2}, \quad r_2 = \frac{b_1 + b_3}{2}, \quad r_3 = \frac{b_2 + b_3}{2}, \tag{87}$$

$r_3 \geq r_2 \geq r_1$, the system (86) assumes the diagonal (Riemann) form

$$\frac{\partial r_j}{\partial T} + V_j(r_1, r_2, r_3)\frac{\partial r_j}{\partial X} = 0, \quad j = 1, 2, 3, \tag{88}$$

where the characteristic velocities $V_3 \geq V_2 \geq V_1$ are expressed as certain combinations of the complete elliptic integrals of the first and the second kind. No summation over the repeated indices is assumed in (88). The variables r_j are called *Riemann invariants*. It is known very well (see, for instance, Whitham 1974) that Riemann invariants can always be found for systems consisting of two quasilinear equations, but for systems of three or more equations they generally do not exist. The remarkable fact of the existence of the Riemann invariants for the KdV modulation system (86) is connected with the preservation of integrability under the averaging. We will discuss this issue in the next section.

Although the direct derivation of the system (88) from the conservative system (86) is a rather laborious task (see, for instance, Kamchatnov 2000), one can easily obtain explicit expressions for the characteristic velocities by taking advantage of the established existence of the diagonal form (88) and the known dependence of the cnoidal wave parameters (7)–(9) on the Riemann invariants r_j via equation (87). For that, we notice that for the wavenumber conservation law (78) to be consistent with the diagonal system (88), the following relationships must hold (Gurevich et al. 1991, 1992; see also Kamchatnov 2000)

$$V_j = \frac{\partial(kc)}{\partial r_j} \bigg/ \frac{\partial k}{\partial r_j}, \quad j = 1, 2, 3, \tag{89}$$

where k, m and c specified by equations (7)–(9) are expressed in terms of the Riemann invariants (87) as

$$k = \frac{\pi(r_3 - r_1)^{1/2}}{K(m)}, \quad m = \frac{r_2 - r_1}{r_3 - r_1}, \quad c = 2(r_1 + r_2 + r_3). \tag{90}$$

Indeed, introducing in (78) the Riemann invariants explicitly and using (88) we obtain

$$\sum_{j=1}^{3} \left\{ \frac{\partial \omega}{\partial r_j} - V_j \frac{\partial k}{\partial r_j} \right\} \frac{\partial r_j}{\partial X} = 0. \tag{91}$$

Since the derivatives $\partial_X r_j$ are independent and generally do not vanish, we readily arrive at the 'potential' representation (89), which can be viewed as a generalisation to a nonlinear case of the group velocity notion defined in linear wave theory as $c_g = \partial \omega / \partial k$. Substituting (90) into (89), we obtain explicit expressions for the characteristic velocities in terms of the complete elliptic integrals,

$$V_1 = 2(r_1 + r_2 + r_3) - 4(r_2 - r_1)\frac{K(m)}{K(m) - E(m)}, \tag{92}$$

$$V_2 = 2(r_1 + r_2 + r_3) - 4(r_2 - r_1)\frac{(1-m)K(m)}{E(m) - (1-m)K(m)}, \tag{93}$$

$$V_3 = 2(r_1 + r_2 + r_3) + 4(r_3 - r_1)\frac{(1-m)K(m)}{E(m)}. \tag{94}$$

We now study the Whitham equations in two distinguished asymptotic limits: linear ($m \to 0$) and soliton ($m \to 1$). For that, we write down the relevant asymptotic expansions of the complete elliptic integrals (see Abramowitz and Stegun 1965):

$$m \ll 1: \quad K(m) = \frac{\pi}{2}\left(1 + \frac{m}{4} + \frac{9}{64}m^2 + \cdots\right),$$

$$E(m) = \frac{\pi}{2}\left(1 - \frac{m}{4} - \frac{3}{64}m^2 + \cdots\right); \tag{95}$$

$$(1 - m) \ll 1: \quad K(m) \approx \frac{1}{2}\ln\frac{16}{1-m},$$

$$E(m) \approx 1 + \frac{1}{4}(1 - m)\left(\ln\frac{16}{1-m} - 1\right). \tag{96}$$

Using the expansions in (95), we get in the harmonic limit $m \to 0$: $V_3 \to 6r_3$; $V_2 \to V_1 \to (12r_1 - 6r_3)$ so that the Whitham system reduces to

$$r_2 = r_1, \quad \frac{\partial r_3}{\partial T} + 6r_3\frac{\partial r_3}{\partial X} = 0, \quad \frac{\partial r_1}{\partial T} + (12r_1 - r_3)\frac{\partial r_1}{\partial X} = 0. \tag{97}$$

In the soliton limit, $m \to 1$ and we have using (96): $V_2 \to V_3 \to (2r_1 + 4r_3)$; $V_1 \to 6r_1$ so that the Whitham system reduces to

$$r_2 = r_3, \quad \frac{\partial r_1}{\partial T} + 6r_1\frac{\partial r_1}{\partial X} = 0, \quad \frac{\partial r_3}{\partial T} + (2r_1 + 4r_3)\frac{\partial r_3}{\partial X} = 0. \tag{98}$$

Thus, the Whitham system (88) admits two *exact* reductions to the systems of lower order via the limiting transitions $r_2 \to r_1$ (linear limit) and $r_2 \to r_3$ (soliton limit). In both limits, one of the Whitham equations converts into the Hopf equation $r_T + 6rr_X = 0$, which coincides with the dispersionless limit of the KdV equation (3), while the remaining two merge into one for the Riemann invariant along a double characteristic. Such a special structure of the KdV–Whitham system (88) will enable us to formulate and solve some physically important boundary-value problems in Section 5.

4.2 Integrability of the Whitham equations

Since the Whitham system (88) has been obtained by the averaging of the completely integrable KdV equation (3), one can expect that it possesses, apart from the existence of the Riemann invariants, some properties allowing for its exact integration. Indeed, as studies of Tsarev (1985), Krichever (1988) and Dubrovin and Novikov (1989) showed, the integrability is inherited under the Whitham averaging and the Whitham system for the KdV equation is integrable via the so-called *generalised hodograph transform*.

First, one can observe that the Riemann form (88) implies that any $r_j = $ constant is an exact solution of the Whitham equations. We now consider a reduction of the Whitham system when one of the Riemann invariants, say r_3, is constant, $r_3 = r_{30}$. Then for the remaining two $r_{1,2}(X, T)$, one has a 2×2 system, which can be solved using the classical hodograph transform, provided $r_{1X} \neq 0$, $r_{2X} \neq 0$ (see, for instance, Whitham 1974). This is achieved through the change of variables $(r_1, r_2) \mapsto (X, T)$. The resulting (hodograph) system for $X(r_1, r_2)$ and $T(r_1, r_2)$ consists of two *linear* equations:

$$\partial_1 X - V_1(r_1, r_2, r_{30})\partial_1 T = 0, \quad \partial_2 X - V_2(r_1, r_2, r_{30})\partial_2 T = 0, \quad (99)$$

where $\partial_j \equiv \partial/\partial r_j$. Next, we introduce in (99) a substitution

$$W_1(r_1, r_2) = X - V_1 T, \quad W_2(r_1, r_2) = x - V_2 T, \quad (100)$$

to cast it into the form of a symmetric system for $W_{1,2}$:

$$\frac{\partial_1 W_2}{W_1 - W_2} = \frac{\partial_1 V_2}{V_1 - V_2}; \quad \frac{\partial_2 W_1}{W_2 - W_1} = \frac{\partial_2 V_1}{V_2 - V_1}. \quad (101)$$

Now, any solution of the linear system (101) will generate, via (100), a *local* solution $\{r_1(X, T), r_2(X, T), r_{30}\}$ of the Whitham system. One can see that analogous systems can be obtained for any two pairs of the Riemann invariants, provided the third invariant is constant. In 1985, Tsarev showed that even if *all three* Riemann invariants vary, any smooth non-constant solution of the Whitham systems (88) can be obtained from the algebraic system

$$X - V_j(r_1, r_2, r_3)T = W_j(r_1, r_2, r_3), \quad i = 1, 2, 3, \quad (102)$$

where the functions W_j are found from the overdetermined system of linear partial differential equations,

$$\frac{\partial_i W_j}{W_i - W_j} = \frac{\partial_i V_j}{V_i - V_j}, \quad i, j = 1, 2, 3, \ i \neq j. \quad (103)$$

Now, the condition of integrability of the nonlinear diagonal system (88) is reduced to the condition of consistency for the overdetermined linear system (103), which has the form (see Tsarev 1985; Dubrovin and Novikov 1989)

$$\partial_i \left(\frac{\partial_j V_k}{V_j - V_k} \right) = \partial_j \left(\frac{\partial_i V_k}{V_i - V_k} \right), \quad i \neq j, \ i \neq k, \ j \neq k. \quad (104)$$

It is not difficult to show that the characteristic velocities (92)–(94) satisfy relationships (104) and thus, the KdV–Whitham system (88) is integrable. The construction in (102)–(104) is known as the generalised hodograph transform.

4.3 Whitham equations and spectral problem

There exists a deep connection between the Whitham equations (88) and the spectral problem associated with the original KdV equation. This connection has been discovered and thoroughly studied in the paper by Flaschka, Forest and McLaughlin (FFM)(1979). Let us consider the cnoidal wave solution (10), taken with negative sign, as a potential in the linear Schrödinger equation (26) in the associated spectral problem. It is well known that the spectrum of the periodic Schrödinger operator generally consists of an *infinite number* of disjoint intervals called *bands*. Correspondingly, the 'forbidden' zones between the bands are called *gaps*. The unique property of the cnoidal wave solution (10) is that its spectrum contains only *one* finite band. To be exact, the spectral set for the potential $-u_{cn}(x)$ is $S = \{\lambda : \lambda \in [\lambda_1, \lambda_2] \bigcup [\lambda_3, \infty)\}$. This fact had been known long before the creation of the soliton theory in connection with the so-called *Lamè* potentials. The soliton studies showed that the cnoidal wave solutions of the KdV equation represent the simplest case of potentials belonging to a general class of so-called *finite-gap potentials* discovered by Novikov (1974) and Lax (1975). These finite-gap potentials can be expressed in terms of the Riemann theta functions and give rise to *multiphase* almost periodic solutions of the KdV equation.

It is clear that the cnoidal wave solution can be parametrised by three spectral parameters λ_1, λ_2 and λ_3 instead of the roots of the polynomial, b_1, b_2 and b_3 (see equation (87)). The remarkable general fact established by FFM is that the Riemann invariants of the Whitham system (88) coincide with the endpoints of the spectral bands of finite-gap potential. In particular, for the single-gap solution (the cnoidal wave), $r_1 = \lambda_1, r_1 = \lambda_1, r_3 = \lambda_3$. Thus, the spectral problem provides one with the most convenient set of modulation parameters (the Riemann invariants) and, therefore, the Whitham equations (88) describe slow evolution of the *spectrum* of multiphase KdV solutions. The general theory of finite-gap integration and the spectral theory of the Whitham equations are quite technical. However, in the case of the single-phase waves, which is the most important from the viewpoint of fluid dynamics applications, a simple universal method has been developed by Kamchatnov (2000) enabling one to construct periodic solutions and the Whitham equations directly in Riemann invariants for a broad class of integrable nonlinear dispersive wave equations.

5 Undular bores

5.1 Formation and structure of an undular bore

Let the initial data $u(x, 0) = u_0(x)$ for the KdV equation (3) have the form of a smooth step with a single inflection point at the origin:

$$u_0(-\infty) = u^-, \quad u_0(+\infty) = u^+, \quad u_0'(x) < 0, \quad u_0''(0) = 0. \tag{105}$$

Let $\epsilon = |u_0'(0)|/(u^- - u^+) \ll 1$, i.e. the characteristic width of the transition region is much larger than the characteristic wavelength, which is unity in the KdV equation (3). The qualitative picture of the KdV evolution of such a large-scale step transition is as follows. During the initial stage of the evolution, $|u_x| \sim \epsilon$, $|u_{xxx}| \sim \epsilon^3$, hence $|u_{xxx}| \ll |uu_x|$, and one can neglect the dispersive term in the KdV equation. The evolution at this stage is approximately described by the dispersionless (classical) limit of the KdV equation,

$$u \approx r(x, t) : \quad r_t + 6rr_x = 0, \quad r(x, 0) = u_0(x), \tag{106}$$

which is the Hopf equation for a simple wave (see Landau and Lifshitz 1987; Whitham 1974). The evolution (106) leads to a gradient catastrophe, which occurs at the inflection point at a certain *breaking time* $t \to t_b, x \to x_b: r_x \to -\infty, r_{xx} \to 0$. In classical (dissipative) hydrodynamics the wave breaking leads to the formation of a *shock*, which can be asymptotically represented as a discontinuity where intense dissipation occurs and the flow parameters undergo a rapid change (see Whitham 1974). Instead, in dispersive hydrodynamics the resolution of the breaking singularity occurs through the generation of nonlinear waves. Indeed, for $t > t_b$ (without loss of generality we can put $t_b = 0$), one can no longer neglect the KdV dispersive term u_{xxx} in the vicinity of the breaking point and, as a result, the regularisation of the singularity happens through the generation of nonlinear oscillations confined to a finite, albeit expanding, space region. This oscillatory structure represents a dispersive analog of a shock wave and is often called *an undular bore*. In nature, the undular bore is a phenomenon occurring in some rivers (River Severn in England and River Dordogne in France are among the best known) and representing a wave-like transition between two basic flows with different depth. Although the mathematical modelling of such shallow-water undular bores requires taking into account weak dissipation (Benjamin and Lighthill 1954; Johnson 1970; Whitham 1974), which stabilises the expansion of the oscillatory zone, it is now customary to use the term 'undular bore' for any wave-like transition between two different smooth flows in solutions of nonlinear dispersive systems. The significance of the study of purely conservative, unsteady undular bores is twofold: they can be viewed as an initial stage of the development of undular bores with small dissipation to a steady state and also, importantly, they represent a universal mechanism of soliton generation out of non-oscillatory initial or boundary conditions in conservative systems. In fact, the study of purely dispersive undular bores has stimulated a number of important discoveries in modern nonlinear wave theory.

An undular bore solution to the KdV equation has the distinctive spatial structure, which has been first observed in numerical simulations. Near the leading edge of the undular bore, the oscillations appear to be close to successive solitons, while in the vicinity of the trailing edge they are nearly linear (see Fig. 1). Using knowledge of

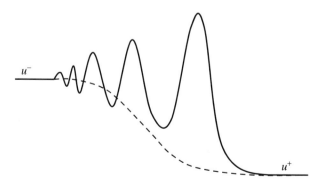

Figure 1: Qualitative structure of the undular bore evolving from the initial step (dashed line).

this qualitative structure, Gurevich and Pitaevskii (GP) (1974) made an assumption that the undular bore (it is called a 'collisionless shock wave' in the original GP paper) represents a *modulated single-phase solution* of the KdV equation. More precisely, the undular bore is a constructed cnoidal wave solution (10) where the parameters b_1, b_2 and b_3 (with an account of relations (87)) evolve according to the Whitham equations (88). We note that this whole asymptotic construction has been recovered in a later rigorous theory of the singular zero-dispersion limit of the KdV equation by Lax, Levermore and Venakides 1994 (see Section 4), but in the GP approach it is a plausible assumption.

5.2 Gurevich–Pitaevskii problem: general formulation

In the GP approach, the accent is shifted from finding the exact solution of the KdV equation (3) with the initial conditions (105) to finding the corresponding exact solution of the associated Whitham system (88). Hence, the first task is to translate the initial data $u_0(x)$ of the KdV equation into some initial or boundary conditions for the Whitham system. The Whitham equations (88), unlike the KdV equation itself, are of hydrodynamic type, i.e. they do not contain higher order derivatives. This, similar to classical ideal hydrodynamics, results in the non-existence of the global solution for the general initial-value problem (see, for instance, Whitham 1974; Landau and Lifshitz 1987). Indeed, due to nonlinearity of the Whitham equations, the modulation parameters r_1, r_2 and r_3 would develop infinite derivatives on their profiles in finite time, which would make the whole Whitham system invalid, as it is based on the assumption of the *slowly* modulated cnoidal wave. Therefore, the Whitham equations should be supplied with the boundary conditions ensuring the existence of the global solution. It is physically natural to require the continuous matching of the mean flow $\bar{u}(x, t)$ in the undular bore with the smooth flow $u(x, t)$ outside the undular bore at some free boundaries. Also, it follows from the (assumed) structure of the undular bore that the matching of the undular bore with the non-oscillating external flow must occur at the points of the linear ($m \to 0$) degeneration of the undular bore at the trailing edge and soliton ($m \to 1$) degeneration at the leading edge. The mean value given by formula (83) can be readily expressed in terms of the Riemann invariants r_j using the relationships (87). Then using asymptotic expansions (95) and (96) of the complete elliptic integrals in the linear and soliton limits, we obtain

$$\bar{u}|_{m=0} = r_3, \quad \bar{u}|_{m=1} = r_1. \tag{107}$$

Now the problem of the continuous matching of the mean flow can be formulated in the following, mathematically accurate, way. From here on, we will not introduce slow variables $X = \epsilon x$, $T = \epsilon t$ in the modulation equations (88) explicitly, instead, we assume that the small parameter ϵ naturally arises *in the solution* of the KdV equation as the ratio of the characteristic wavelength to the width of the oscillations zone. Let the upper (x, t) half-plane be split into three domains: $\{x \in \mathbb{R}, t > 0\} = \{(-\infty, x^-(t)) \cup [x^-(t), x^+(t)] \cup (x^+(t), +\infty)\}$ (see Fig. 2), in which the solution is governed by different equations: outside the interval $[x^-(t), x^+(t)]$ it is governed

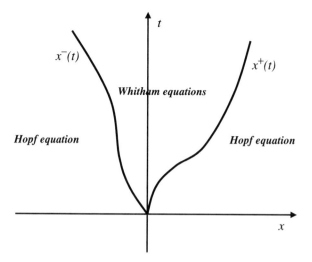

Figure 2: Splitting of the (x, t) - plane in the Gurevich–Pitaevskii problem.

by the Hopf equation (106), while within the interval $[x^-(t), x^+(t)]$ the dynamics is described by the Whitham equations (88) so that the following matching conditions are satisfied:

$$x = x^-(t): \quad r_2 = r_1, \ r_3 = r,$$
$$x = x^+(t): \quad r_2 = r_3, \ r_1 = r, \tag{108}$$

where $r(x, t)$ is the solution of the Hopf equation (106) and the boundaries $x^\pm(t)$ are unknown at the onset. One can see that conditions (108) are consistent with the limiting structure of the Whitham equations given by equations (97) and (98) and thus, the lines $x^\pm(t)$ represent free boundaries. It follows from equation (108), and the limiting properties of the Whitham velocities described in Section 4.1, that these boundaries are defined by the multiple characteristics of the Whitham system for $m = 0$ $(x = x^-(t))$ and $m = 1$ $(x = x^+(t))$ and are found from the ordinary differential equations

$$dx^-/dt = (12r_1 - 6r_3)|_{x=x^-}, \quad dx^+/dt = (2r_1 + 4r_3)|_{x=x^+} \tag{109}$$

defined *on the solution* $\{r_j(x, t)\}$ of the GP problem.

5.3 Decay of an initial discontinuity

As an important example, where a simple representation of the undular bore can be obtained, we consider the decay of an initial discontinuity problem. We take the initial data in the form of a sharp step,

$$t = 0: \quad u = A > 0 \text{ if } x < 0, \quad u = 0 \text{ for } x > 0, \tag{110}$$

which implies the immediate formation of an undular bore. We now assume the modulation description of the undular bore and make use of the GP problem formulation. First we observe that, since both initial data (110) and the modulation equations (88) are invariant with respect to the linear transformation $x \to Cx$, $t \to Ct$, $C = $ constant, the modulation variables must be functions of a self-similar variable $s = x/t$ alone so that $r_j = r_j(s)$. Thus, the Whitham system (88) reduces to the system of ODEs:

$$\frac{dr_j}{ds}(V_j - s) = 0, \quad j = 1, 2, 3. \tag{111}$$

The GP matching conditions (108) then assume the form

$$
\begin{aligned}
s = s^-: \quad & r_2 = r_1, \quad r_3 = A, \\
s = s^+: \quad & r_2 = r_3, \quad r_1 = 0,
\end{aligned}
\tag{112}
$$

where $s^\pm$ are the (unknown) speeds of the undular bore edges, $x^\pm = s^\pm t$. The boundary-value problem (111), (112) has the solution in the form of a centred simple wave in which all but one Riemann invariants are constant:

$$r_1 = 0, \quad r_3 = A, \quad V_2(0, r_2, A) = s, \tag{113}$$

or, explicitly, using the expression (93) for $V_2(r_1, r_2, r_3)$,

$$2A\left\{1 + m - \frac{2(1-m)m}{E(m)/K(m) - (1-m)}\right\} = \frac{x}{t}, \tag{114}$$

where $m = (r_2 - r_1)/(r_3 - r_1) = r_2/A$. The obtained solution for the Riemann invariants is schematically shown in Fig. 3. It provides the required modulation of the cnoidal wave (10) in the undular bore transition. Since the solution (114) represents a characteristic fan, it never breaks for $t > 0$ and therefore is global.

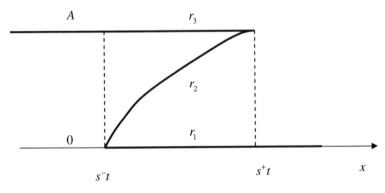

Figure 3: Riemann invariant behaviour in the self-similar undular bore.

The speeds $s^{\mp}$ of the trailing and leading edges of the undular bore are found by assuming $m = 0$ and $m = 1$, respectively, in the solution (114):

$$s^- = s(0) = -6A, \quad s^+ = s(1) = 4A. \tag{115}$$

Thus, the undular bore is confined to an expanding zone $-6At \leq x \leq 4At$. The amplitude of the lead soliton is simply $a^+ = 2(r_3 - r_1) = 2A$. This agrees with the semi-classical IST result (70), which gives the amplitude of the greatest soliton evolving out of the large-scale initial perturbation with the amplitude $u_{0\,max} = A$. Indeed, solution (114) of the decay of a step problem can be viewed as intermediate asymptotics for $1 \ll t \ll l$ in the problem of the evolution of a spatially extended rectangular profile of the width $l \gg 1$: $u_0(x) = A$ if $x \in [-l, 0]$ and $u_0(x) = 0$, otherwise. It can also be readily inferred from (114) that the phase velocity $c = 2(r_1 + r_2 + r_3) = 2A(1 + m) > V_2(0, r_2, A)$ if $m < 1$ and $c = V_2(0, r_2, A)$ for $m = 1$. Thus, any individual crest within the wavetrain moves towards the leading edge of the undular bore, i.e. for any crest $m \to 1$ as $t \to \infty$. In this sense, the undular bore evolves into a soliton train.

Using asymptotic expansions (95) and (96) for the complete elliptic integrals, we obtain the asymptotic behaviour of the modulus m near the undular bore boundaries. Near the trailing edge, we have

$$m \simeq (s - s^-)/9 \ll 1, \tag{116}$$

which also describes the amplitude variations since $m = a/A$ for the solution under study. Near the leading edge, we get with logarithmic accuracy:

$$1 - m \simeq (s^+ - s)/2 \ln(1/(s^+ - s)) \ll 1, \tag{117}$$

which, in particular, yields the asymptotic behaviour

$$\bar{u} \simeq 12k, \quad k \simeq 2\pi/\ln(1/(s^+ - s)) \tag{118}$$

for the mean value and the wavenumber respectively.

It should be stressed that, although formula (114) represents an exact solution of the Whitham equations, the full undular bore solution (i.e. the travelling wave (10) modulated by (114)) is an *asymptotic solution* of the KdV equation as the Whitham method itself is based on the perturbation theory. As a result, the location of the undular bore is determined up to the typical wavelength (indeed, the initial phase x_0 in (10) is lost after the Whitham averaging) and thus, the accuracy ϵ of the obtained undular bore description can be estimated as the ratio of the typical soliton width, which is $\mathcal{O}(1)$, to the width of the oscillations zone, $l \cong 10At$, i.e. $\epsilon \sim t^{-1}$. The obtained undular bore description is thus asymptotically accurate as $t \to \infty$.

If the constant A in the initial conditions (110) is negative, $A < 0$, the initial step does not break and the undular bore is not generated. Instead, the asymptotic solution of the KdV equation, in this case, is a rarefaction wave,

$$u = 0 \text{ if } x > 0, \quad u = x/6t \text{ if } 6At < x < 0, \quad u = 0 \text{ if } x < 6At. \tag{119}$$

The solution (119) contains weak discontinuities at $x = 0$ and $x = 6At < 0$. These discontinuities are resolved in the full solution of the KdV equation with the small-amplitude linear wavetrains which smooth out with time (see Gurveich and Pitaevskii 1974, for further details).

Fornberg and Whitham (1978) compared the modulation solution of Gurevich and Pitaevskii with the full numerical solution of the KdV equation with the step initial conditions and found a very good agreement between the two. More recently, Apel (2003) utilised the GP solution for modelling internal undular bores in the Strait of Gibraltar in the Mediterranean Sea.

5.4 General solution of the Gurevich–Pitaevskii problem

In the case when the initial data is not the step function, the similarity solution (114) is not applicable and to construct the appropriate solution of the GP problem, one needs a more general solution of the Whitham equations, in which two or all three Riemann invariants vary. Such solutions can be constructed using the generalised hodograph transform described in Section 4.2. However, although the resulting hodograph equations (103) are linear, they are still too complicated to be treated directly. It was shown by Gurevich et al. (1991, 1992) that the scalar substitution (cf. (89))

$$W_i = \frac{\partial_i(kf)}{\partial_i k}, \quad i = 1, 2, 3, \tag{120}$$

where $k(r_1, r_2, r_3)$ is given by (90), and $f(r_1, r_2, r_3)$ – the unknown function – is compatible with system (103) and reduces it to the system of classical Euler–Poisson–Darboux equations,

$$2(r_i - r_j)\partial_{ij}^2 f = \partial_i f - \partial_j f, \quad i, j = 1, 2, 3, \ i \neq j. \tag{121}$$

It is not difficult to show that the overdetermined system (121) is consistent and has a general solution (Eisenhart 1919)

$$f = \sum_{i=1}^{3} \int \frac{\phi_i(\tau)d\tau}{\sqrt{(r_3 - \tau)(r_2 - \tau)(\tau - r_1)}}, \tag{122}$$

where $\phi_i(\tau)$ are arbitrary (generally complex) functions. Next, by applying the GP matching conditions (108) to equations (102), (120) and (122), the unknown functions ϕ_i can be expressed in terms of the linear Abel transforms of the monotone parts of the KdV initial profile $u_0(x)$. Then, for monotonically decreasing the initial data $u(x, 0) = u_0(x)$, $u_0'(x) < 0$ with a single breaking point at $u = 0$, the resulting solution for the function $f(\mathbf{r})$ assumes the form (Gurevich et al. 1992).

$$f = \frac{1}{\pi(r_3 - r_2)^{1/2}} \int_{r_2}^{r_3} \frac{W(\tau)}{\sqrt{\tau - r_1}} K(z)d\tau + \frac{1}{\pi(r_2 - r_1)^{1/2}} \int_{r_1}^{r_2} \frac{W(\tau)}{\sqrt{r_3 - \tau}} K(z^{-1})d\tau,$$

$$\tag{123}$$

where

$$z = \left[\frac{(r_2 - r_1)(r_3 - \tau)}{(r_3 - r_2)(\tau - r_1)} \right]^{1/2} \qquad (124)$$

and $W(u) = u_0^{-1}(x)$ is the inverse function of the initial profile. The sought dependence $r_j(x, t)$ in the undular bore is now found by the substitution of the solution (123) into formulae (120) and (102) and resolving them for $r_j(x, t)$. It was shown by Krylov *et al.* (1992) that, for decaying initial functions $u_0(x)$, the longtime asymptotics of the solution of the GP problem agrees with the semi-classical Karpman formula (69), thus providing another connection of the Whitham theory with the IST method.

6 Propagation of KdV soliton through a variable environment

In many physical situations, the properties of the medium vary in space. The weakly nonlinear wave propagation in such media is described by the KdV equation (1) with the variable coefficients $\alpha(t)$ and $\beta(t)$ (see, for instance, Johnson 1997; Grimshaw 2001). One should note that in the modelling of the variable environment effects, the variables x and t in the KdV equation (1) are not necessarily the same physical space and time co-ordinates as in the traditional interpretation of the KdV equation (3). However, for convenience, we shall retain the same x, t notations here. Assuming $\alpha \neq 0$, we introduce the new variables

$$t' = \frac{1}{6} \int_0^t \alpha(\hat{t}) d\hat{t}, \quad \lambda(t') = \frac{6\beta}{\alpha}, \qquad (125)$$

so that, on omitting the superscript for t, equation (1) becomes

$$u_t + 6uu_x + \lambda(t)u_{xxx} = 0. \qquad (126)$$

The physical problems modelled by (126) include shallow-water waves moving over an uneven bottom, internal gravity waves in lakes of varying cross sections long waves in rotating fluids contained in cylindrical tubes and many others (see, for instance, Johnson 1997; Grimshaw 2001 and references therein).

We shall assume the following physically reasonable behaviour for the variable coefficient $\lambda(t)$: let $\lambda(t)$ be constant, say 1 for $t < 0$, then changes smoothly until $t = t_1$ and then again is a (different) constant λ_1. Suppose that a soliton solution of the constant-coefficient KdV equation (3) (which incidentally is a solution of equation (126) for $t < 0$) is moving through the medium so that it reaches the point $x = 0$ at $t = 0$. It is clear that for $t > 0$ it is no longer an exact solution of equation (126) and some wave modification must occur. Generally, equation (126) is not integrable by the IST method, so the problem should be solved numerically. There are, however, two distinct limiting cases that can be treated analytically.

If the medium properties change rapidly, i.e. $0 < t_1 \ll 1$, then the known soliton waveform (13) can be used as an initial condition for the KdV equation (126) with $\lambda = \lambda_1 = $ constant (see Johnson 1997). This problem can be solved by the IST method and the outcome is that for $\lambda_1 > 0$ the initial soliton *fissions* into N solitons and some radiation (see Section 3.7). If $\lambda_1 < 0$, then the initial soliton (13) completely transforms into the linear radiation (see Section 3.6).

An opposite situation occurs when the medium properties vary slowly so that $\sigma \sim |\lambda_t| \ll 1$, i.e. $t_1 \sim \sigma^{-1} \gg 1$. In this case, one can use an adiabatic approximation to describe the solitary wave variations to leading order. Thus, we assume that λ is slowly varying so that

$$\lambda = \lambda(T), \quad T = \sigma t, \quad \sigma \ll 1. \tag{127}$$

Then, the asymptotic expansion of the slowly varying solitary wave is given by

$$u = u_0 + \sigma u_1 + \cdots , \tag{128}$$

where the leading term is given by

$$u_0 = a \operatorname{sech}^2 \left\{ \gamma \left(x - \frac{\Phi(T)}{\sigma} \right) \right\}, \tag{129}$$

so that

$$d\Phi/dT = c = 2a = 4\lambda\gamma^2. \tag{130}$$

Note that for $\lambda = $ constant, formulae (129) and (130) reduce to the standard KdV soliton expression (13) with $b_1 = 0$.

The variations of the amplitude a, the inverse half-width parameter γ and the speed c with the slow time variable T are determined by noticing that the variable-coefficient KdV equation (126) possesses the momentum conservation law

$$\int_{-\infty}^{\infty} u^2 dx = \text{constant}. \tag{131}$$

Substitution of (129) into (131) readily shows that

$$\frac{\gamma}{\gamma_0} = \left(\frac{\lambda_0}{\lambda} \right)^{2/3}, \tag{132}$$

where the subscript '0' indicates quantities evaluated at $T = 0$, i.e. $\lambda_0 = 1$.

It follows from (129), (130) and (132) that the slowly varying solitary wave, u_0, is now completely determined. However, the variable-coefficient KdV equation (126) also has a conservation law for 'mass'

$$\int_{-\infty}^{\infty} u dx = \text{constant}, \tag{133}$$

which is not satisfied by the leading order adiabatic expression (129). The situation can be remedied by taking into account the next term in the asymptotic expansion (128) and allowing $\int_{-\infty}^{\infty} u_1(x)dx$ to be $\mathcal{O}(\sigma^{-1})$.

More precisely, conservation of mass is assured by the generation of a trailing shelf u_s, such that $u = u_0 + u_s$, where u_s typically has an amplitude $\mathcal{O}(\sigma)$ and is supported on the interval $0 < x < \Phi(T)/\sigma$ (see Newell 1985, Chapter 3 and the references therein; or Grimshaw and Mitsudera 1993). Thus, the shelf stretches over a zone of $\mathcal{O}(\sigma^{-1})$ and hence carries $\mathcal{O}(1)$ mass. The law (133) for conservation of mass then shows that

$$\int_{-\infty}^{\Phi/\sigma} u_s \, dx + \int_{-\infty}^{\infty} u_0 \, dx = \text{constant.} \tag{134}$$

The second term on the left-hand side of (134) is readily found to be

$$\frac{2a}{\gamma} = 4\lambda\gamma = 4\gamma_0 \lambda^{1/3}, \tag{135}$$

on using (129), (130) and (132) in turn. Next, we assume that $u_s = \sigma q(X, T)$, where

$$X = \sigma x, \quad T = \sigma t. \tag{136}$$

Since the spatial overlap between u_0 and u_s is small compared with the shelf width, one can assume $u \approx u_s$ for $0 < x < \Phi(T)/\sigma$ and, therefore, $u_s(x, t)$ must satisfy (126). Then we have

$$q_T + 6\sigma q q_X + \sigma^2 \lambda(T) q_{XXX} = 0. \tag{137}$$

For the sake of definiteness we shall assume that $\lambda > 0$. Next, on differentiating (134) with respect to T, we find that to leading order in σ,

$$q = -\frac{2}{\gamma_0} \lambda_T \lambda^{-1/3} \equiv Q(T) \quad \text{at } X = \Phi(T). \tag{138}$$

Now, (137) and (138) together with (130) and (132) present a completely formulated boundary-value problem.

We assume that $\lambda(T)$ is a smooth function and, in particular, has a continuous second derivative so that $Q(T)$ has a continuous first derivative. Then at least in some finite neighbourhood of the curve $X = \Phi(T)$, one can neglect the dispersive term in the KdV equation (137) and approximately describe the evolution by the Hopf equation

$$q_T + 6\sigma q q_X = 0 \tag{139}$$

with the same boundary condition (138). The solution to the boundary-value problem (139), (138) is readily found using the characteristics:

$$q = Q(T_0), \quad X - \Phi(T_0) = 6\sigma Q(T_0)(T - T_0), \tag{140}$$

where T_0 is a parameter along the initial curve $\Phi(T)$. The expression (140) remains valid until neighbouring characteristics intersect and a breaking of the $q(x)$-profile begins. This occurs when

$$\Phi'(T_0) + 6\sigma Q'(T_0)(T - T_0) - 6\sigma Q(T_0) = 0. \tag{141}$$

Since $\Phi' = c > 0$ and $\sigma \ll 1$, it follows that the singularity forms in finite time, only if $Q'(T_0) < 0$ at least for some values of T_0. Let T_b be the minimum value, as T_0 varies, such that (141) is satisfied. Then the breaking singularity forms first at T_b and the corresponding value X_b is determined from equation (140). It can be easily seen from equation (141) that $T_b = \mathcal{O}(\sigma^{-1})$ provided $Q'(T_0) = \mathcal{O}(1)$.

It is clear that in the vicinity of the breaking point (X_b, T_b), the Hopf equation (139) no longer describes the evolution of q adequately and the full KdV equation (137) should be considered. As a result, the singularity is regularised by an unsteady

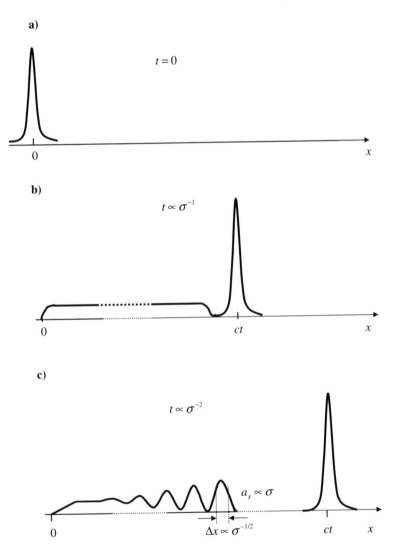

Figure 4: Formation and evolution of a soliton trailing shelf undular bore (a) KdV soliton at $t = 0$; (b) formation of a constant-amplitude $a \sim \sigma$ elevation shelf at $t \sim \sigma^{-1}$; (c) generation of the undular bore in the shelf at $t \sim \sigma^{-2}$.

undular bore (see Fig. 4). One can see that the undular bore forms at times $T \sim T_b \sim \sigma^{-1}$, which are much greater than the time $T_1 = \sigma t_1 = \mathcal{O}(1)$ for the change of the variable coefficient $\lambda(T)$ from 1 to λ_1. Thus, the trailing shelf undular bore essentially forms and evolves in the region where $\lambda(T) = \lambda_1 > 0$ is constant and, as a result, the problem of the longtime trailing shelf evolution reduces to solving the familiar constant-coefficient KdV equation,

$$q_T + 6\sigma q q_X + \sigma^2 \lambda_1 q_{XXX} = 0. \tag{142}$$

The initial data for equation (142), $q_0(X) = q(X, 0)$ is obtained from the boundary condition (138), specified at the initial curve $\Phi(T)$, by its projection, along the characteristics of the Hopf equation (139), onto X-axis. The sign of $q_0(X)$ depends on the sign of λ_T in (138) (i.e. on whether λ_1 is greater or less than unity). If $\lambda_1 < 1$, then $q_0(X) > 0$. As the variations of the function $Q(T)$ in (138) are determined by the variations of the function $\lambda(T)$, the characteristic spatial scale l of this equivalent initial condition $q_0(X)$ is $\mathcal{O}(1)$. On the other hand, the typical wavelength of the travelling wave solutions of equation (142) is $\Delta X \approx \sigma^{1/2} \ll l$. Simple rescaling to the standard form (3) shows that the condition (57) of applicability of the semi-classical asymptotics in the IST method is satisfied. Therefore, if $\lambda_1 < 1$, the trailing shelf eventually decomposes into a large number of small-amplitude solitons and one can use Karpman's formula (69) to obtain the amplitude distribution in the soliton train. Alternatively, if $\lambda_1 > 1$, then $q_0(X) < 0$ and the trailing shelf converts as $t \to \infty$ into a linear wave packet, again, via an intermediate stage of an undular bore.

The undular bore stage of the trailing shelf evolution has been studied in detail in El and Grimshaw (2002), where it has been argued that for the variable environment with $\lambda_T < 0$, the solitons generated in the trailing shelf undular bore can be identified with the secondary solitons in the fissioning scenario. Thus, both extreme cases of soliton propagation through a variable environment are now reconciled by showing that the trailing shelf contains the seeds for the generation of secondary solitons, albeit on a long time scale.

References

Ablowitz, M.J. & Segur, H., *Solitons and the Inverse Scattering Transform*, SIAM: Philadelphia, 1981.

Abramowitz, M. & Stegun, I.A. (eds.), *Handbook of Mathematical Functions*, Dover: New York, 1965.

Apel, J.P., A new analytical model for internal solitons in the ocean. *Journ. Phys. Oceanogr.*, **33**, pp. 2247–2269, 2003.

Benjamin, T.B. & Lighthill, M.J., On cnoidal waves and bores. *Proc. Roy. Soc.*, **A224**, pp. 448–460, 1954.

Dodd, R.K., Eilbeck, J.C., Gibbon, J.D. & Morris, H.C., *Solitons and Nonlinear Wave Equations*, Academic Press: London, 1982.

Drazin, P.G. & Johnson, R.S., *Solitons: An Introduction*, Cambridge University Press: Cambridge, 1989.

Dubrovin, B.A. & Novikov, S.P., Hydrodynamics of weakly deformed soliton lattices. Differential geometry and Hamiltonian theory. *Russian Math. Surveys*, **44**, pp. 35–124, 1989.

Eisenhart, L.P., Triply conjugate systems with equal point invariants. *Ann. Math.* (2), **20**, pp. 262–273, 1919.

El, G.A. & Grimshaw, R.J.H., Generation of undular bores in the shelves of slowly varying solitary waves. *Chaos*, **12**, pp. 1015–1026, 2002.

Flaschka, H., Forest, G. & McLaughlin, D.W., Multiphase averaging and the inverse spectral solutions of the Korteweg–de Vries equation. *Comm. Pure Appl. Math.*, **33**, pp. 739–784, 1979.

Fornberg, B. & Whitham, G.B., A numerical and theoretical study of certain nonlinear wave phenomena. *Phil. Trans. R. Soc. Lond.*, **A289**, pp. 373–404, 1978.

Gardner, C.S., Greene, J.M., Kruskal, M.D. & Miura, R.M., Method for solving the Korteweg–de Vries equation. *Phys. Rev. Lett.*, **19**, pp. 1095–1097, 1967.

Grimshaw, R., Slowly varying solitary waves. I – Korteweg–de Vries equation. *Proc. Roy. Soc.*, **A368**, pp. 359–375, 1979.

Grimshaw, R., Internal solitary waves (Chapter 1). *Environmental Stratified Flows*, Kluwer: Boston, pp. 1–28, 2001.

Grimshaw, R. & Mitsudera, H., Slowly varying solitary wave solutions of the perturbed Korteweg–de Vries equation revisited. *Stud. Appl. Math.*, **90**, pp. 75–86, 1993.

Gurevich, A.V. & Pitaevsky, L.P., Nonstationary structure of a collisionless shock wave. *Sov. Phys. JETP*, **38**, pp. 291–297, 1974.

Gurevich, A.V., Krylov, A.L. & El, G.A., Riemann wave breaking in dispersive hydrodynamics. *JETP Lett.*, **54**, pp. 102–107, 1991.

Gurevich, A.V., Krylov, A.L. & El, G.A., Evolution of a Riemann wave in dispersive hydrodynamics. *Sov. Phys. JETP*, **74**, pp. 957–962, 1992.

Johnson, R.S., A non-linear equation incorporating damping and dispersion. *J. Fluid Mech.*, **42**, pp. 49–60, 1970.

Johnson, R.S., *A Modern Introduction to the Mathematical Theory of Water Waves*, Cambridge University Press: Cambridge, 1997.

Kamchatnov, A.M., *Nonlinear Periodic Waves and Their Modulations – An Introductory Course*, World Scientific: Singapore, 2000.

Karpman, V.I., An asymptotic solution of the Korteweg–de Vries equation. *Phys. Lett. A*, **25**, pp. 708–709, 1967.

Karpman, V.I., *Nonlinear Waves in Dispersive Media*. Pergamon: Oxford, 1975.

Krichever, I.M., The method of averaging for two-dimensional 'integrable' equations. *Funct. Anal. Appl.*, **22**, pp. 200–213, 1988.

Krylov, A.L., Khodorovskii, V.V. & El, G.A., Evolution of nonmonotonic perturbation in Korteweg–de Vries hydrodynamics. *JETP Letters*, **56**, pp. 323–327, 1992.

Landau, L.D. & Lifshitz, E.M., *Quantum Mechanics: Nonrelativistic Theory*, 3rd edn, Pergamon: Oxford, 1977.

Landau, L.D. & Lifshitz, E.M., *Fluid Mechanics*, 4th edn, Pergamon: Oxford, 1987.

Lax, P.D., Integrals of nonlinear equations of evolution and solitary waves. *Commun. Pure Appl. Math.*, **21**, pp. 467–490, 1968.

Lax, P.D., Periodic solutions of the KdV equation. *Commun. Pure Appl. Math.*, **28**, pp. 141–188, 1975.

Lax, P.D., Levermore, C.D. & Venakides, S., The generation and propagation of oscillations in dispersive initial value problems and their limiting behavior. *Important Developments in Soliton Theory*, eds. A.S. Focas & V.E. Zakharov, Springer-Verlag: New York, pp. 205–241, 1994.

Luke, J.C., A perturbation method for nonlinear dispersive wave problems. *Proc. Roy. Soc.*, **A292**, pp. 403–412, 1966.

Nayfeh, A.H., *Introduction to Perturbation Techniques*, Wiley: New York, 1981.

Newell, A.C., *Solitons in Mathematics and Physics*, SIAM: Philadelphia, 1985.

Novikov, S.P., The periodic problem for the Korteweg–de Vries equation. *Functional Anal. Appl.*, **8**, pp. 236–246, 1974.

Novikov, S.P., Manakov, S.V., Pitaevskii, L.P. & Zakharov, V.E., *The Theory of Solitons: The Inverse Scattering Method*, Consultants Bureau: New York, 1984.

Scott, A.C., *Nonlinear Science: Emergence and Dynamics of Coherent Structures*, Oxford University Press: Oxford, 2003.

Tsarev, S.P., On Poisson brackets and one-dimensional systems of hydrodynamic type. *Soviet Math. Dokl.*, **31**, pp. 488–491, 1985.

Whitham, G.B., Non-linear dispersive waves. *Proc. Roy. Soc.*, **A283**, pp. 238–291, 1965.

Whitham, G.B., *Linear and Nonlinear Waves*, Wiley: New York, 1974.

Zabusky, N.J. & Kruskal, M.D., Interactions of "solitons" in a collisionless plasma and the recurrence of initial states. *Phys. Rev. Lett.*, **15**, pp. 240–243, 1965.

CHAPTER 3

Solitary waves in water: numerical methods and results

J.-M. Vanden-Broeck
School of Mathematics, The University of East Anglia, Norwich, UK.

Abstract

We describe accurate numerical methods to compute solitary waves of arbitrary amplitude. The effects of gravity and surface tension are included in the dynamic boundary condition. Both two- and three-dimensional solutions are considered. It is found that the structure of the different families of solutions is complex. Therefore, we use appropriate asymptotic solutions to identify the ranges of values of the parameters where solutions of a particular type can be expected. Solitary waves, solitary waves with decaying oscillatory tails and generalised solitary waves are presented. Extensions to free surface flows with the effects of vorticity and electric fields included, are also reviewed.

1 Introduction

In this chapter, we consider periodic and solitary waves propagating at the surface of a layer of fluid bounded below by a horizontal bottom and above by a free surface. We assume that the fluid is inviscid and that the flow is irrotational. We neglect the motion of the fluid on top of the free surface and assume that it is characterised by a constant pressure. This is a good approximation for a layer of water bounded above by the atmosphere. We shall consider both two- and three-dimensional problems. The two-dimensional problem can be motivated by considering the two-dimensional free surface flow past an obstacle at the bottom of a channel (see Fig. 1). Here the obstacle is half of a circle.

This two-dimensional configuration provides a good approximation for the three-dimensional free surface flow past a long cylinder perpendicular to the plane of the figure (except near the ends of the cylinder). The cross section of the cylinder is the half of a circle shown in Fig. 1. We assume that the acceleration due to gravity g is acting in the negative vertical direction.

Two-dimensional free surface flows, such as the one shown in Fig. 1, often have waves on their free surfaces. When dissipation is neglected, these waves approach uniform trains of waves in the far field. Therefore, a fundamental problem in the theory of free surface flows is the study of a uniform train of two-dimensional waves of wavelength λ, extending from $x = -\infty$ to $x = \infty$ and travelling at a constant velocity c. The flow configuration is illustrated in Fig. 2.

Of particular interest here is the limit $\lambda \to \infty$. This can generate solitary waves, i.e. non-periodic travelling waves approaching a flat free surface in the far field (see Fig. 3). When surface tension is included in the dynamic boundary condition,

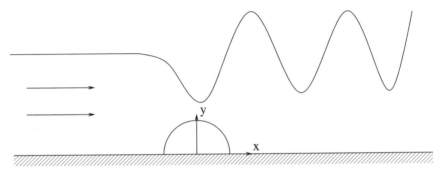

Figure 1: The two-dimensional free surface flow past a submerged circle.

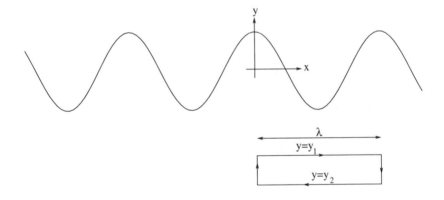

Figure 2: A two-dimensional train of waves viewed in a frame of reference moving with the wave. The free surface profile has wavelength λ. The fluid is bounded below by a horizontal bottom of equation $y = -h$. Also shown is the rectangular contour used in (9).

there are also solitary waves with oscillatory tails. Some of them have decaying oscillations and are therefore true solitary waves with a flat free surface in the far field (see Chapters 1 and 7). There are both waves of elevation and of depression (see Figs 4 and 5).

Others have oscillations of constant amplitudes in the far field and are referred to as generalised solitary waves to distinguish them from true solitary waves (see Fig. 6 and Chapter 1).

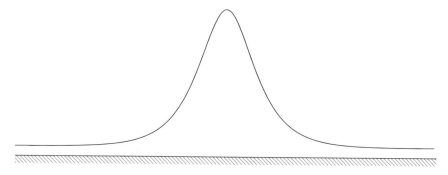

Figure 3: A solitary wave.

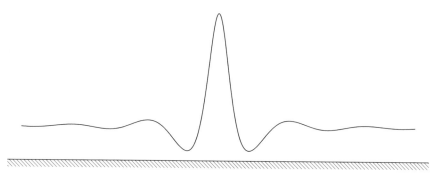

Figure 4: An elevation solitary wave with a decaying oscillatory tail.

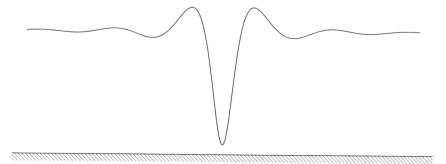

Figure 5: A depression solitary wave with a decaying oscillatory tail.

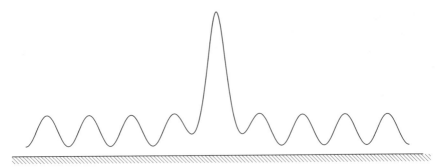

Figure 6: A generalised solitary wave.

In the following sections, we will describe numerical schemes to compute both periodic and solitary waves. The schemes for periodic waves provide approximations for solitary waves when larger and larger values of the wavelength are chosen.

As we shall see, there are also solitary waves in three dimensions. These waves are characterised by decaying oscillations in the direction of propagation and monotonic decay in the direction perpendicular to that of the propagation. Some aspects of these waves are also discussed in Chapter 1.

2 Formulation

We formulate the problem of Fig. 2 by assuming that the fluid is inviscid and incompressible and that the flow is irrotational (although some effects of vorticity are discussed in Section 7). We choose a frame of reference moving with the wave so that the flow is steady. We can then write the velocity in the fluid as the gradient of a potential function $\phi(x, y)$, such that the horizontal and vertical components of the velocity are given by $u = \phi_x$ and $v = \phi_y$, respectively. We formulate the problem as

$$\phi_{xx} + \phi_{yy} = 0, \quad -h < y < \eta(x), \tag{1}$$

$$\phi_y = \phi_x \eta_x, \quad \text{on } y = \eta(x), \tag{2}$$

$$\phi_y = 0, \quad \text{on } y = -h, \tag{3}$$

$$\frac{1}{2}(\phi_x^2 + \phi_y^2) + gy - \frac{T}{\rho} \frac{\eta_{xx}}{(1 + \eta_x^2)^{3/2}} = B, \quad \text{on } y = \eta(x), \tag{4}$$

$$\nabla\phi(x + \lambda, y) = \nabla\phi(x, y), \tag{5}$$

$$\eta(x + \lambda) = \eta(x), \tag{6}$$

$$\int_0^\lambda \eta(x)dx = 0, \tag{7}$$

$$\frac{1}{\lambda}\int_0^\lambda \phi_x dx = c, \quad y = \text{constant}. \tag{8}$$

Here, g is the acceleration due to gravity (assumed to act in the negative y-direction), T is the surface tension, B is the Bernoulli constant, ρ is the density of the fluid, $y = -h$ is the equation of the bottom and $y = \eta(x)$ is the equation of the (unknown) free surface. Equations (2) and (3) are the kinematic boundary conditions on the free surface and on the bottom respectively. Equation (4) is the dynamic boundary condition on the free surface. Relations (5) and (6) are the periodicity conditions, which require the solution to be periodic with a wavelength λ. Equation (7) fixes the origin of the y coordinates as the mean water level. Finally, (8) defines the velocity c as the average value of $u = \phi_x$ at a level $y = \text{constant}$ in the fluid. The value of c is independent of the value $y = \text{constant}$ chosen. This can be shown by applying Stokes' theorem to the vector velocity (u, v) with a contour C consisting of two horizontal lines $y = y_1$ and $y = y_2$ and two vertical lines separated by a wavelength (see Fig. 2). Since the flow is irrotational, Stokes' theorem implies

$$\int_C u dx + v dy = 0. \tag{9}$$

The contributions from the two vertical lines cancel by periodicity and (9) gives

$$\int_0^\lambda [u]_{y=y_1} dx = \int_0^\lambda [u]_{y=y_2} dx. \tag{10}$$

Since y_1 and y_2 are arbitrary, the integral on the left-hand side of (8) is independent of the level $y = \text{constant}$ chosen in the fluid.

The system (1)–(8) is referred to as the water wave equations because it models waves travelling at the interface between water and air (although it applies also to other fluids).

3 Linear solutions

A trivial solution of the system (1)–(8) is

$$\phi = cx, \quad \eta(x) = 0 \quad \text{and} \quad B = \frac{c^2}{2}. \tag{11}$$

Linear waves are obtained by seeking a solution as a small perturbation of the exact solution (11). Therefore, we write

$$\phi(x, y) = cx + \varphi(x, y) \tag{12}$$

and assume that both $\varphi(x, y)$ and $\eta(x)$ are small. Substituting (12) into (1)–(8) and dropping the nonlinear terms in φ and η, we obtain the linear system

$$\varphi_{xx} + \varphi_{yy} = 0, \quad -h < y < 0, \tag{13}$$

$$\varphi_y = c\eta_x = 0, \quad y = 0, \tag{14}$$

$$\varphi_y = 0, \quad y = -h, \tag{15}$$

$$-\frac{T}{\rho}\eta_{xx} + c\varphi_x + g\eta = 0, \quad y = 0, \tag{16}$$

$$\nabla\varphi(x + \lambda, y) = \nabla\varphi(x, y), \tag{17}$$

$$\eta(x + \lambda) = \eta(x), \tag{18}$$

$$\int_0^\lambda \eta(x)dx = 0, \tag{19}$$

$$\frac{1}{\lambda}\int_0^\lambda \varphi_x dx = 0, \quad y = \text{constant}. \tag{20}$$

We choose the origin of x at a crest and assume that the wave is symmetric about $y = 0$. Thus we impose

$$\varphi(-x, y) = -\varphi(x, y), \tag{21}$$

$$\eta(-x) = \eta(x). \tag{22}$$

Applying the method of separation of variables to (13) and using the condition (15) give

$$\varphi(x, y) = \sum_{n=1}^{\infty} B_n \cosh nk(y + h) \sin nkx, \tag{23}$$

where B_n are constants. The periodicity condition (18) and the symmetry condition (22) imply that $\eta(x)$ can be represented by the Fourier series

$$\eta(x) = A_0 + \sum_{n=1}^{\infty} A_n \cos nkx, \tag{24}$$

where A_n are also constants and $k = 2\pi/\lambda$ is the wavenumber. The conditions (19) and (14) give $A_0 = 0$ and

$$cA_n = -B_n \sinh nkh, \quad n = 1, 2, 3, \ldots. \tag{25}$$

Substituting (23) and (24) into (16) yields

$$\frac{T}{\rho}A_n n^2 k^2 + gA_n + cB_n nk \cosh nkh = 0, \quad n = 1, 2, 3 \ldots. \tag{26}$$

Eliminating B_n between (25) and (26) yields

$$\left(g + \frac{T}{\rho} n^2 k^2 - \frac{c^2 nk}{\sinh nkh} \cosh nkh \right) A_n = 0, \quad n = 1, 2, 3, \ldots . \quad (27)$$

Since we seek a non-trivial periodic solution $\eta(x) \neq 0$, we can assume without loss of generality that $A_1 \neq 0$, then (27) with $n = 1$ implies

$$c^2 = \left(\frac{g}{k} + \frac{T}{\rho} k \right) \tanh kh. \quad (28)$$

Relation (27) for $n > 1$ implies

$$A_n = 0, \quad n = 2, 3, \ldots, \quad (29)$$

provided

$$g + \frac{T}{\rho} n^2 k^2 - \frac{c^2 nk}{\sinh nkh} \cosh nkh \neq 0, \quad n = 2, 3, \ldots . \quad (30)$$

If (30) is satisfied, the solution of the linear problem is

$$\varphi = -\frac{cA_1}{\sinh kh} \cosh k(y + h) \sin kx, \quad (31)$$

$$\eta = A_1 \cos kx. \quad (32)$$

If the condition (30) is not satisfied for some integer value m of n, i.e. if

$$g + \frac{T}{\rho} m^2 k^2 - \frac{c^2 mk}{\sinh mkh} \cosh mkh = 0, \quad (33)$$

the solution is

$$\varphi = -\frac{cA_1}{\sinh kh} \cosh k(y + h) \sin kx - \frac{cA_m}{\sinh mkh} \cosh mk(y + h) \sin mkx, \quad (34)$$

$$\eta = A_1 \cos kx + A_m \cos mkx, \quad (35)$$

where A_m is an arbitrary constant. In the theory of linear waves, it is simply assumed that $A_m = 0$. However $A_m \neq 0$, when developing nonlinear theories (i.e. when improving the linear approximation (12) by adding nonlinear corrections or when computing numerically fully nonlinear solutions). It is also found that there are several possible values for A_m (see Wilton 1915; Nayfeh 1970; Schwartz and Vanden-Broeck 1979; Chen and Saffman 1979, 1980b; Hogan 1980 and others). For example, in water of infinite depth, $A_2 = \pm A_1/2$ when $m = 2$. In other words, there are two solutions when $m = 2$. Further analytical and numerical work indicates that there are many different families of gravity–capillary periodic waves.

The corresponding free surface profiles do not usually decay monotonically from a crest to a trough but have 'dimples'.

The velocity c is called the phase velocity and (28) the (linear) dispersion relation. It is convenient to rewrite (28) in the dimensionless form:

$$F^2 = \left(\frac{1}{kh} + \tau kh \right) \tanh kh, \tag{36}$$

where

$$F = \frac{c}{(gh)^{1/2}} \tag{37}$$

is the Froude number and

$$\tau = \frac{T}{\rho gh^2} \tag{38}$$

is the Bond number. Graphs of F^2 versus $1/(kh) = \lambda/(2\pi h)$ for four values of τ are shown in Fig. 7.

The curves illustrate that F^2 is a monotonically decreasing function of λ/h when $\tau > 1/3$ and has a minimum for $\tau < 1/3$. As $\lambda/h \to \infty$, $F \to 1$. The different behaviours for $\tau < 1/3$ (minimum) and $\tau > 1/3$ (monotone decay) in Fig. 7 have many implications in particular when studying nonlinear periodic and solitary gravity–capillary waves. In particular, the condition (33) expresses that waves of wavenumbers k and mk travel at the same speed. This can occur only if the curves

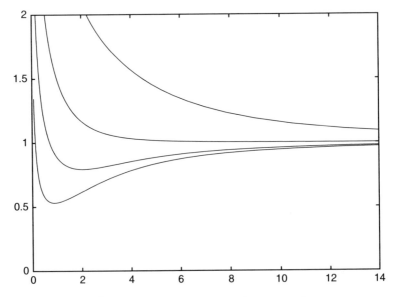

Figure 7: Values of F^2 versus $1/(kh)$. The curves from top to bottom correspond to $\tau = 1.3$, $\tau = 1/3$, $\tau = 0.1$ and $\tau = 0.05$. For $\tau < 1/3$, the curves have a minimum, whereas for $\tau > 1/3$ they decrease monotonically.

of Fig. 7 have a minimum, i.e. if $\tau < 1/3$. If $\tau > 1/3$, the condition (33) cannot be satisfied and the solution of the problem is unique.

Long periodic waves for $\tau < 1/3$ were calculated numerically by Hunter and Vanden-Broeck (1983a), Vanden-Broeck (1991) and Champneys *et al.* (2002). The numerical procedure is described in Section 5. Analytical results were derived by Hunter and Scherule (1988), Iooss and Kirchgassner (1990, 1992), Beale (1991), Sun (1991) and others (see Dias and Khariff 1999 for a review). The numerical results show that the multiple branches of periodic solutions approach generalised solitary waves as $\lambda/h \to \infty$. This can easily be explained by looking at the sketch of Fig. 6 as representing one wavelength of a typical long wave on a branch of periodic waves. This branch is then characterised by 9 crests and 10 troughs on one wavelength of its profile. The next branch of periodic waves obtained by increasing λ/h is then characterised by 11 crests and 10 troughs on one wavelength. The next branch would have 11 crests and 12 troughs on one wavelength of its profile. In other words, moving from one branch to the next adds two crests or two troughs. In the limit as $\lambda/h \to \infty$, this generates a generalised solitary wave characterised by an infinite number of crests and troughs. For a given value of τ, the generalised solitary waves form a two-parameter family of solutions. The ripples in the tail of the generalised solitary waves are questionable physically because they occur on both sides and therefore do not satisfy the radiation condition. Therefore, an important question is whether one of the two parameters can be chosen so that the amplitude of the ripples in the far field vanishes. One of the features of these ripples is that their amplitude is an exponentially small function of $F - 1$, as has been shown by exponential asymptotics (Sun and Shen 1993) and by rigorous application of centre manifold and normal form theory (Lombardi 2000). The question has been rigorously answered in the negative for τ sufficiently close to 1/3 by Sun (1999). Champneys et al. (2002) showed numerically that the answer is in fact negative for $9/50 < \tau < 1/3$. Similar results were found by Turner and Vanden-Broeck (1992), Akylas and Grimshaw (1992), Sha and Vanden-Broeck (1997) and Michallet and Dias (1999) for interfacial and internal waves.

The existence of branches of solitary waves with decaying oscillatory tails (see Figs 4 and 5) is also related to the minima in Fig. 7 for $\tau < 1/3$. This can be explained intuitively as follows. The profiles of Figs 4 and 5 look like waves whose amplitudes slowly vary. We should then expect these waves of varying amplitude to travel at the phase velocity while their envelopes travel at the group velocity. Since phase and group velocity have in general different values, the waves of Figs 4 and 5 cannot be expected to be steady unless the group and phase velocity are equal. This is exactly what happens at the minima of Fig. 7. This explains intuitively why the branches of solitary waves with decaying tails bifurcate from the minima in Fig. 7. Numerical calculations of such waves were provided by Hunter and Vanden-Broeck (1983a), Zufuria (1987), Longuet-Higgins (1989), Vanden-Broeck and Dias (1992), Dias *et al.* (1996) and others. Existence proofs were given by Iooss and Kirchgassner (1990) and others (see Dias and Khariff 1999 for a review). Analytical approximations were obtained by Dias and Iooss (1993), Akylas (1993) and Longuet-Higgins (1993).

4 Weakly nonlinear results

A natural way to improve on the linear theory of Section 3 is to assume expansions of the form

$$\phi(x, y) = cx + \epsilon\phi_1(x, y) + \epsilon^2\phi_2(x, y) + \epsilon^3\phi_3(x, y) + O(\epsilon^4), \tag{39}$$

$$\eta(x) = \epsilon\eta_1(x) + \epsilon^2\eta_2(x) + \epsilon^3\eta_3(x) + O(\epsilon^4), \tag{40}$$

$$c = c_0 + \epsilon c_1 + \epsilon^2 c_2 + \epsilon^3 c_3 + O(\epsilon^4), \tag{41}$$

$$B = B_0 + \epsilon B_1 + \epsilon^2 B_2 + \epsilon^3 B_3 + O(\epsilon^4). \tag{42}$$

Here, ϵ is a measure of the amplitude of the wave. The terms $\epsilon\phi_1(x, y)$ and $\epsilon\eta_1(x)$ are the linear approximations (31), (32) (if (30) is satisfied) and the terms of order $\epsilon^2, \epsilon^3, \ldots$ are nonlinear corrections. The first few terms in the expansions (39)–(42) have been calculated by Stokes (1847) for pure gravity waves (i.e. waves with $T = 0$). Schwartz (1974) developed a systematic way to compute a large number of terms and obtained accurate nonlinear solutions by summing the expansions with Padé approximants. This work was extended by Longuet-Higgins (1975) and by Cokelet (1977). Similar calculations could be performed for pure capillary waves (i.e. waves with $g = 0$). However, Crapper (1957) and Kinnersley (1976) showed that the system (1)–(8) with $g = 0$ has exact solutions. For gravity–capillary waves ($g \neq 0$ and $T \neq 0$), the weakly nonlinear solutions can be calculated by assuming that $\phi_1(x, y)$, $\eta_1(x)$ and c_0^2 are defined by the right-hand sides of (31), (32) and (28). However, it is found that the higher-order corrections are unbounded when the condition (33) is satisfied. Appropriate expansions when (33) is satisfied can be derived by assuming that $\phi_1(x, y)$, $\eta_1(x)$ and c_0^2 are given by the right-hand sides of (34), (35) and (28). As indicated in Section 3, this leads to several values of A_m. Explicit calculations for $m = 2$ and $m = 3$ were performed by Wilton (1915) and Nayfeh (1970). Chen and Saffman (1979) also computed solutions that were valid when the right-hand side of (30) is small. This reveals the existence of many different families of solutions. Analytical calculations for larger values of m become rapidly prohibitive and numerical solutions have to be sought, for example, by using the numerical codes described in Section 5 (alternative methods along the work of Schwartz (1974) were developed by Hogan (1980)).

A limitation of the expansions (39)–(42) is that the ratio of successive terms becomes unbounded as $h \to 0$ (e.g. $|\epsilon\phi_2(x, y)/\phi_1(x, y)| \to \infty$ as $h \to 0$). Therefore, the expansions (39)–(42) are non-uniform as $h \to 0$. To obtain uniform expansions valid for small values of h, it is necessary to assume that both the depth h and the amplitude of the waves are small. This can be described as follows.

First we introduce the amplitude parameter

$$\alpha = \frac{a}{h} \tag{43}$$

and the depth parameter

$$\beta = \frac{h^2}{l^2}. \tag{44}$$

Here, a is a typical amplitude and l a typical horizontal length scale. We write the elevation of the free surface as $y = h + \eta$, where h is the undisturbed elevation. We also introduce the velocity scale $c_0 = (gh)^{1/2}$.

Next, we rescale the variables as

$$\tilde{x} = \frac{x}{l}, \quad \tilde{y} = \frac{y}{h}, \quad \tilde{t} = \frac{c_0}{l}t, \tag{45}$$

$$\tilde{\eta} = \frac{\eta}{a}, \quad \tilde{\phi} = \frac{c_0}{gla}\phi. \tag{46}$$

The different scalings in the x and y direction in (45) is crucial in deriving shallow-water approximations.

We rewrite the nonlinear equations (1)–(4) in terms of the new variables (45) and (46). This leads after dropping the tilde

$$\beta\phi_{xx} + \phi_{yy} = 0, \quad 0 < y < 1 + \alpha\eta, \tag{47}$$

$$\phi_y = 0, \quad \text{on } y = 0, \tag{48}$$

$$\alpha\phi_x\eta_x - \frac{1}{\beta}\phi_y = 0, \quad \text{on } y = 1 + \alpha\eta, \tag{49}$$

$$\eta + \frac{1}{2}\alpha\phi_x^2 + \frac{1}{2}\frac{\alpha}{\beta}\phi_y^2 - \tau\beta\frac{\eta_{xx}}{(1 + \alpha^2\eta_x^2)^{3/2}} = 0, \quad \text{on } y = 1 + \alpha\eta. \tag{50}$$

We seek solutions for periodic waves of wavelength λ and introduce the dimensionless wavelength

$$\tilde{l} = \frac{\lambda}{l}. \tag{51}$$

The Froude number F is defined by

$$F = \frac{c}{(gh)^{1/2}} = \frac{\alpha}{\tilde{l}}\int_0^{\tilde{l}}\phi_x\,\mathrm{d}x. \tag{52}$$

The shallow-water equations are derived by assuming β to be small and α of order unity. These equations do not have travelling waves in the absence of surface tension because the dispersive effects are neglected. The inclusion of dispersive effects into the shallow water theory leads to the Korteweg–de Vries equation (Korteweg and de Vries 1895). This equation can be derived by assuming β to be small and α of order β. Thus, we let

$$\alpha = \beta = \epsilon. \tag{53}$$

We seek steady solutions in a frame of reference moving with the wave and expand η, ϕ, F and B in powers of ϵ as

$$\eta = \eta_0 + \epsilon\eta_1 + \epsilon^2\eta_2 + \cdots , \tag{54}$$

$$\phi = \frac{Bx}{\epsilon} + \phi_0 + \epsilon\phi_1 + \epsilon^2\phi_2 + \cdots , \tag{55}$$

$$B = B_0 + \epsilon B_1 + \epsilon^2 B_2 + \cdots , \tag{56}$$

$$F = F_0 + \epsilon F_1 + \epsilon^2 F_2 + \cdots . \tag{57}$$

Substituting (54)–(57) into (47)–(50) leads after some algebra

$$B_0 = F_0 = 1, \tag{58}$$

$$\phi_0' = -\eta_0, \tag{59}$$

$$2F_1\eta_0' - 3\eta_0\eta_0' + \left(\tau - \frac{1}{3}\right)\eta_0''' = 0. \tag{60}$$

The primes in (59) and (60) denote derivatives with respect to x. Equation (60) is the steady Korteweg–de Vries equation. The unsteady version is derived and discussed in Chapter 1.

Korteweg and de Vries (1895) showed that periodic solutions of (60) can be found in the closed form. As the wavelength $\tilde{l}$ tends to infinity, these waves tend to the solitary wave solution

$$\eta_0 = a\,\text{sech}^2\frac{x}{b}, \tag{61}$$

$$a = 2F_1, \quad b = \left[\frac{4(1 - 3\tau)}{3a}\right]^{1/2}. \tag{62}$$

When $\tau < 1/3$, these are elevation waves with values of the Froude number greater than 1 (see Fig. 3). When $\tau > 1/3$, they are depression waves with a Froude number less than 1. Equation (62) shows that the maximum slope of the wave profile becomes large for τ near $1/3$ and that the solution ceases to exist altogether for $\tau = 1/3$. Therefore, the Korteweg–de Vries equation (60) is not valid for τ close to $1/3$. This is due to the fact that the dispersive effects disappear from (60) as $\tau \to 1/3$.

To derive a weakly nonlinear equation valid near $\tau = 1/3$, we take

$$\alpha = \epsilon^2, \quad \beta = \epsilon, \tag{63}$$

in (47)–(50). Then we expand η, ϕ, B, F and τ as

$$\eta = \eta_0 + \epsilon \eta_1 + \epsilon^2 \eta_2 + \cdots, \tag{64}$$

$$\phi = \frac{Bx}{\epsilon^2} + \phi_0 + \epsilon \phi_1 + \epsilon^2 \phi_2 + \epsilon^3 \phi_3 + \cdots, \tag{65}$$

$$\tau = \frac{1}{3} + \epsilon \tau_1 + \epsilon^2 \tau_2 + \cdots, \tag{66}$$

$$B = B_0 + \epsilon B_1 + \epsilon^2 B_2 + \cdots, \tag{67}$$

$$F = F_0 + \epsilon F_1 + \epsilon^2 F_2 + \cdots. \tag{68}$$

Substituting (63)–(68) into (47)–(50), we obtain after some algebra

$$F_0 = B_0 = 1, \quad F_1 = B_1 = 0, \quad F_2 = B_2, \tag{69}$$

$$\eta_0 = -\phi_0', \tag{70}$$

$$2F_2 \eta_0' - 3\eta_0 \eta_0' + \tau_1 \eta_0''' - \frac{1}{45} \eta_0^v = 0. \tag{71}$$

Equation (71) provides a generalisation of the Korteweg–de Vries equation for τ close to $1/3$. It is often referred to as the fifth-order Korteweg–de Vries equation. For the unsteady version of (71), see, for example, Hunter and Scherule (1988) and Chapter 1.

In the next sections, we describe numerical methods to compute fully nonlinear solutions. The resulting numerical results extend the weakly nonlinear results discussed in this section to the fully nonlinear regime. An important finding is that the elevation solitary waves (61) predicted by the Korteweg–de Vries equation (60), when $\tau < 1/3$, are inaccurate in the sense that the solutions of the nonlinear equations do not have flat free profiles in the far field like in Fig. 3, but are generalised solitary waves (see Fig. 6). The fully nonlinear solutions obtained by the numerical procedures described in the next section provide the correct generalised solitary waves. The fifth-order Korteweg–de Vries also predicts them correctly when their amplitude is small and τ is close to $1/3$. Some of the properties of these generalised solitary waves have already been discussed in Section 3.

5 Numerical methods for periodic waves

We consider again the train of periodic waves shown in Fig. 2. The waves have wavelength λ and propagate to the left in a channel bounded below by a horizontal bottom. We take a frame of reference moving with the wave and seek steady solutions. For the numerical computations, we choose cartesian coordinates with $x = y = 0$ at a crest of the wave. Gravity is acting in the negative y-direction (see Fig. 8).

We introduce in addition to the potential function ϕ, a stream function ψ. We choose $\phi = 0$ at $x = y = 0$ and $\psi = 0$ on the free surface. We denote by $-Q$ the

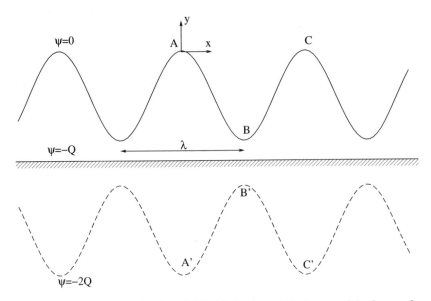

Figure 8: A wave propagating in a fluid of finite depth. The image of the free surface
in the bottom is also shown.

constant value of ψ on the horizontal bottom. We also denote by u and v the
horizontal and vertical components of the velocity.

As in Section 2, we define the phase velocity c as the average horizontal velocity
at a constant level of y in the fluid. Using (8) we find that

$$c = \frac{1}{\lambda} \int_0^\lambda u(x, y) \mathrm{d}x = \frac{1}{\lambda} \int_0^\lambda \phi_x \mathrm{d}x = \frac{1}{\lambda}[\phi(\lambda, 0) - \phi(0, 0)] = \frac{1}{\lambda}\phi(\lambda, 0). \quad (72)$$

Therefore, $\phi(\lambda, 0) = c\lambda$ and the assumed periodicity and symmetry of the wave
implies that

$$\phi\left(\frac{n\lambda}{2}, 0\right) = c\frac{n\lambda}{2}, \quad n = 0, \pm 1, \pm 2, \pm 3, \dots . \quad (73)$$

Following Stokes, we use the potential function and the stream function as independent variables. We define the complex potential

$$f = \phi + i\psi. \quad (74)$$

We satisfy the kinematic boundary condition on the horizontal bottom by reflecting
the flow into the bottom (see Fig. 8). The resulting flow in the complex potential
plane is shown in Fig. 9. It is the strip $-\infty < \phi < \infty$, $-2Q < \psi < 0$. The obvious
advantage of working in the complex f-plane is that the unknown free surface of
Fig. 8 has been mapped onto the known boundary $\psi = 0$ of Fig. 9.

We define the function $\alpha + i\beta$ by the relation

$$\alpha + i\beta = \frac{1}{u - iv} = x_\phi + iy_\phi. \quad (75)$$

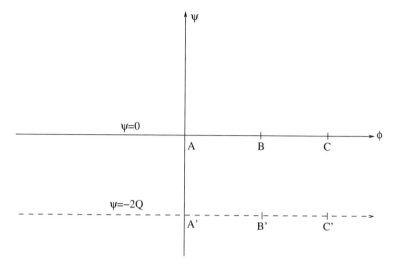

Figure 9: The flow of Fig. 8 in the complex potential plane.

Since $u - iv$ is an analytic function of f, $\alpha + i\beta$ is also an analytic function of f.
Next we map the f-plane into the unit disk in the t-plane by the transformation

$$t = e^{-\frac{2i\pi f}{c\lambda}} = e^{-\frac{2i\pi \phi}{c\lambda}} e^{\frac{2\pi \psi}{c\lambda}} . \tag{76}$$

The flow configuration in the t-plane is shown in Fig. 10.

The free surface has been mapped on the circle $|t| = 1$, the bottom into the circle $|t| = e^{-\frac{2\pi Q}{c\lambda}}$ and the image of the free surface into the circle $|t| = e^{-\frac{4\pi Q}{c\lambda}}$.

We need to express the fact that $\alpha + i\beta$ is an analytic function of t in the annulus $r_0^2 < |t| < 1$, where

$$r_0 = e^{-\frac{2\pi Q}{c\lambda}} . \tag{77}$$

There are two ways to do it. The first is to represent $\alpha + i\beta$ as a Laurent expansion in powers of t. This leads to a numerical procedure which we call series truncation (see Section 5.1). The second is to apply Cauchy integral formula to the function $\alpha + i\beta$ in the annulus. This leads to the boundary integral equation method (see Section 5.2).

5.1 Series truncation methods

We represent $\alpha + i\beta$ by the Laurent expansion,

$$\alpha + i\beta = a_0 + \sum_{n=1}^{\infty} a_n t^n + \sum_{n=1}^{\infty} b_n t^{-n} = a_0 + \sum_{n=1}^{\infty} a_n e^{-\frac{2i\pi n f}{c\lambda}} + \sum_{n=1}^{\infty} b_n e^{\frac{2i\pi n f}{c\lambda}} . \tag{78}$$

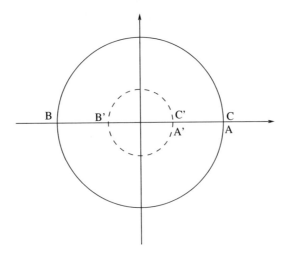

Figure 10: The flow of Fig. 8 in the complex t-plane.

Since

$$\alpha(\phi, -2Q) + i\beta(\phi, -2Q) = \alpha(\phi, 0) - i\beta(\phi, 0), \tag{79}$$

it follows from (78) that

$$b_n = a_n r_0^{2n}, \quad n = 1, 2, \ldots . \tag{80}$$

Furthermore (73) implies that

$$a_0 = \frac{1}{c}. \tag{81}$$

Substituting (80) and (81) into (78) gives

$$\alpha + i\beta = \frac{1}{c} + \sum_{n=1}^{\infty} a_n e^{-\frac{2i\pi n f}{c\lambda}} + \sum_{n=1}^{\infty} a_n r_0^{2n} e^{\frac{2i\pi n f}{c\lambda}}. \tag{82}$$

We shall find the coefficients a_n in (82) such that the dynamic boundary condition (4) is satisfied. To do so, we need to rewrite (4) in terms of α and β. Using (75) we obtain after some algebra

$$\frac{1}{2}\frac{1}{\tilde{\alpha}^2 + \tilde{\beta}^2} + g\int_0^\phi \tilde{\beta}(\varphi)d\varphi - \frac{T}{\rho}\frac{\tilde{\alpha}\tilde{\beta}_\phi - \tilde{\alpha}_\phi\tilde{\beta}}{(\tilde{\alpha}^2 + \tilde{\beta}^2)^{3/2}} = B, \tag{83}$$

where $\tilde\alpha(\phi)$ and $\tilde\beta(\phi)$ denote the values of α and β on the free surface $\psi = 0$. In the remaining part of this subsection, we introduce dimensionless variables by using λ as the unit length and c as the unit velocity. Then (83) becomes

$$\frac{1}{2}\frac{1}{\tilde\alpha^2 + \tilde\beta^2} + \frac{2\pi}{\mu}\int_0^\phi \tilde\beta(\varphi)\mathrm{d}\varphi - \frac{\kappa}{2\pi\mu}\frac{\tilde\alpha\tilde\beta_\phi - \tilde\alpha_\phi\tilde\beta}{(\tilde\alpha^2 + \tilde\beta^2)^{3/2}} = B, \qquad (84)$$

where

$$\mu = \frac{2\pi c^2}{g\lambda}, \quad \kappa = \frac{4\pi^2 T}{\rho g\lambda^2}. \qquad (85)$$

We truncate the infinite series in (82) after $N - 1$ terms and determine the $N + 1$ unknowns B, μ and a_n $n = 1, \ldots, N - 1$ by collocation. Thus we introduce the N collocation points

$$\phi_I = \frac{1}{2}\frac{I - 1}{N - 1}, \quad I = 1, \ldots, N, \qquad (86)$$

and satisfy the dynamic boundary condition (84) at these points. This leads to N nonlinear algebraic equations. The last equation is obtained by fixing the amplitude of the wave, for example, by imposing

$$\int_0^{1/2}\tilde\beta(\varphi)\mathrm{d}\varphi = -s, \qquad (87)$$

where s is given. The parameter s is called the steepness of the wave. It is the difference of height between a crest and a trough divided by the wavelength. For given values of s and κ, the system of $N + 1$ equations with $N + 1$ unknowns is solved by iterations (e.g. by using Newton's method).

In the particular case of water of infinite depth, $r_0 = 0$ and (80) implies $b_n = 0$, $n = 1, 2, \ldots$. The Laurent series (78) reduces then to the Taylor series

$$\alpha + i\beta = \sum_{n=0}^\infty a_n t^n. \qquad (88)$$

The representation (88) can be derived directly by noting that (76) maps the flow domain into the unit circle in the t-plane. Therefore $\alpha + i\beta$ can be represented by the Taylor expansion (88).

Many variations of the series truncation method have been proposed. For example, Vanden-Broeck (1986) calculated waves in water of infinite depth by expanding the complex velocity $w = u - iv$ instead of $\alpha + i\beta$ in powers of t, i.e. by writing

$$w = 1 + \sum_{n=1}^\infty c_n e^{-i2\pi nf}. \qquad (89)$$

The series truncation method provides very accurate solutions for pure capillary waves (i.e. $g = 0$). The numerical results are in close agreement with the exact solutions of Crapper (1957) and Kinnersley (1976). In particular, it is found that as the amplitude is increased, the waves ultimately approach configurations with trapped bubbles at their troughs. Extensions of the branches of solutions past these limiting configurations have been proposed by Vanden-Broeck and Keller (1980). Furthermore Blyth and Vanden-Broeck (2004) used the series truncation method to show the existence of new branches of solutions.

Similarly, the series truncation method provides an accurate solution for gravity–capillary waves (see, for example, Chen and Saffman 1980b). Such waves were also calculated by the boundary integral equation method of Section 5.2 (see Schwartz and Vanden-Broeck 1979) and by series summations (see Hogan 1980). The numerical results show the existence of multiple families of solutions described in Section 3. They also show that as the amplitude increases, the waves approach also limiting configurations with trapped bubbles.

For pure gravity waves ($T = 0$), the series truncation method based on (78), (88) or (89) only give accurate solutions for moderate values of steepness. This is attributable to the slow convergence of the expansions as the wave of maximum steepness is approached. The wave of maximum steepness is characterised by a corner at the crests with an enclosed angle of $120°$. This limiting configuration was already anticipated by Stokes but it was not until 1982 that an existence proof was provided (Amick, *et al.* 1982).

Since $t = 1$ at the crests, we have

$$w \approx (1 - t)^{1/3} \quad \text{as } t \to 1, \tag{90}$$

for the wave of maximum steepness. Because of the singularity (90), we cannot expect the expansion (78), (88) and (89) to converge. Also, we can expect these expansions to converge slower and slower as the wave of maximum steepness is approached.

Michell (1883), Olfe and Rottman (1980), Vanden-Broeck (1986) and Vanden-Broeck and Miloh (1995) calculated the wave of maximum steepness as

$$w = (1 - t)^{1/3} \left(1 + \sum_{n=1}^{\infty} d_n e^{-2in\pi f} \right). \tag{91}$$

The expansion (91) satisfies (90) and is therefore convergent. It leads to very accurate results. In particular, it predicts the value $\mu = 1.193072$ for the dimensionless wave speed.

Havelock (1919) and Vanden-Broeck (1986) calculated waves of arbitrary amplitude up to the limiting configuration by expressing w as

$$w = (1 - \beta t)^{1/3} \left(1 + \sum_{n=1}^{\infty} g_n e^{-2in\pi f} \right) \tag{92}$$

and finding the coefficients g_n and β by series truncation and collocation.

By taking the cube of (89), (91) and (92), we see that these expressions are particular cases of

$$w^3 = 1 + \sum_{n=1}^{\infty} f_n e^{-2i\pi nf},$$

(93)

for appropriate choices of the coefficients f_n. Therefore gravity waves of arbitrary amplitude (including the highest) can be calculated by using (93) and collocation. This approach was used by Vanden-Broeck and Miloh (1995).

5.2 Boundary integral equation method

Instead of using the series representations of Section 5.1, we derive a relation between α and β by using Cauchy integral formula.

We apply Cauchy integral formula to the function $\alpha + i\beta$ in the complex t-plane with a contour C chosen as the boundary of the annulus in Fig. 10. This yields

$$\alpha(t_0) + i\beta(t_0) = -\frac{1}{i\pi} \int_C \frac{\alpha(t) + i\beta(t)}{t - t_0} dt.$$

(94)

The contour C consists of the circle $|t| = 1$ oriented clockwise and of the circle $|t| = r_0^2$ oriented anticlockwise. The integral in (94) is a Cauchy principal value.

Using the change of variables (76) (with $\psi = 0$), we rewrite (94) as

$$\tilde{\alpha}(\phi) + i\tilde{\beta}(\phi) = \frac{2}{c\lambda} \int_{\frac{-c\lambda}{2}}^{\frac{c\lambda}{2}} [\tilde{\alpha}(\varphi) + i\tilde{\beta}(\varphi)] \frac{e^{-\frac{2i\pi\varphi}{c\lambda}}}{e^{-\frac{2i\pi\varphi}{c\lambda}} - e^{-\frac{2i\pi\phi}{c\lambda}}} d\varphi$$

$$-\frac{2r_0^2}{c\lambda} \int_{-\frac{c\lambda}{2}}^{\frac{c\lambda}{2}} [\tilde{\alpha}(\varphi) - i\tilde{\beta}(\varphi)] \frac{r_0^2 e^{-\frac{2i\pi\varphi}{c\lambda}}}{r_0^2 e^{-\frac{2i\pi\varphi}{c\lambda}} - e^{-\frac{2i\pi\phi}{c\lambda}}} d\varphi.$$

(95)

Taking the real part of (95), we obtain after some algebra a relation between $\tilde{\alpha}$ and $\tilde{\beta}$. We denote this relation as equation A.

Relation (83) and the equation A define a system of nonlinear integro-differential equations for the unknown functions $\tilde{\alpha}(\phi)$ and $\tilde{\beta}(\phi)$. We solve this system numerically. First we introduce the mesh points

$$\phi_I = -\frac{c\lambda}{2} + \frac{c\lambda}{N-1}(I-1), \quad I = 1, \ldots, N,$$

the midpoints

$$\phi_I^M = \frac{\phi_I + \phi_{I-1}}{2}, \quad I = 1, \ldots, N-1$$

and the unknowns

$$\alpha_I = \tilde{\alpha}(\phi_I), \quad I = 1, \ldots, N$$

and

$$\beta_I = \tilde{\beta}(\phi_I), \quad I = 1, \ldots, N.$$

The integrals in equation A are approximated by the trapezoidal rule (which is spectrally accurate for periodic functions) with $\varphi = \phi_J, J = 1, \ldots, N$ and $\phi = \phi_I^M$, $I = 1, \ldots, N - 1$. The symmetry of the quadrature and of the dicretisation enable us to evaluate the Cauchy principal value as if it were an ordinary integral. We also evaluate the derivatives α_ϕ and β_ϕ by centred difference formulae. The resulting set of nonlinear algebraic equations is solved by Newton's method.

When $r_0 = 0$, the water is of infinite depth and equation A reduces to the simple integral relation

$$\tilde{\alpha}(\phi) = \frac{1}{c} + \frac{1}{c\lambda} \int_{-\frac{c\lambda}{2}}^{\frac{c\lambda}{2}} \tilde{\beta}(\varphi) \cot \frac{\pi(\varphi - \phi)}{c\lambda} \mathrm{d}\varphi. \tag{96}$$

We note that for symmetric waves, the integrals in equation A and (96) can be reduced to integrals from 0 to $c\lambda/2$. The mesh points ϕ_I can then be distributed over half a wavelength.

Explicit details about the numerical procedures can be found in studies by Schwartz and Vanden-Broeck (1979), Vanden-Broeck and Schwartz (1979) and Hunter and Vanden-Broeck (1983a).

Numerical methods were used by Schwartz and Vanden-Broeck (1979) and Vanden-Broeck and Schwartz (1979) to compute steep gravity waves (i.e. waves close to the limiting configuration with a 120° angle at the crest). High accuracy was achieved by concentrating mesh points near the crest. The resulting numerical results are in good agreement with those obtained by the series truncation method of Section 5.1. Chen and Saffman (1980a) and Vanden-Broeck (1983) used a similar numerical procedure to show the existence of new branches of solutions for which successive crests are at different heights. This non-uniqueness is similar to that found recently by Blyth and Vanden-Broeck (2004) for capillary waves (see Section 3).

The boundary integral equation method was also used by Schwartz and Vanden-Broeck (1979) and Hunter and Vanden-Broeck (1983a) to compute periodic gravity–capillary waves. Multiple branches of solutions were found in agreement with the discussion in Section 3.

6 Numerical methods for solitary waves

There are two approaches to investigate solitary waves numerically. The first approach is to derive directly schemes to compute them. The second approach is to compute periodic waves by using the schemes outlined in Section 5 and then to take larger and larger values of the wavelength. The first approach works particularly well for true solitary waves, i.e. non-periodic travelling waves, which approach a flat free surface in the far field (see Figs 3–5). The infinite domain $-\infty < x < \infty$ is then truncated to $-A < x < A$, where A is a large real number and the contribution from $A < x < \infty$ and from $-\infty < x < -A$ are either neglected or approximated by asymptotic solutions. The second approach is more appropriate to compute generalised solitary waves (see Fig. 6). This is because no approximations are made about the behaviour of the flow in the far field, and we can then be confident that the oscillations in the far field are not generated by the truncation of the domain. In the next two subsections, we follow the first approach.

6.1 Boundary integral equation methods

We derive an integral equation formulation for solitary waves by taking the limit $\lambda \to \infty$ in equation A. This leads to

$$
\tilde{\alpha}(\phi) - \frac{1}{c} = -\frac{1}{\pi} \int_{-\infty}^{\infty} \frac{\tilde{\beta}(\varphi)}{\varphi - \phi} d\varphi
$$

$$
+ \frac{1}{\pi} \int_{-\infty}^{\infty} \frac{-\tilde{\beta}(\varphi)(\varphi - \phi) + 2Q(\tilde{\alpha}(\varphi) - 1)}{(\varphi - \phi)^2 + 4Q^2} d\varphi. \tag{97}
$$

Relation (97) and (83) define an integro-differential system for $\tilde{\alpha}$ and $\tilde{\beta}$. It was solved numerically by Hunter and Vanden-Broeck (1983b) for pure gravity–solitary waves. Other similar results were derived by Byatt-Smith and Longuet-Higgins (1976). They show that the values of the Froude number F oscillates infinitely often as the wave of maximum amplitude is approached. This finding is consistent with the asymptotic results of Longuet-Higgins and Fox (1978).

A system similar to (97) and (83) was derived by Vanden-Broeck and Dias (1992) and Dias *et al.* (1996) in water of infinite depth. It was used to compute solitary waves with decaying oscillatory tails (see Figs 4 and 5).

6.2 Series truncation

We describe in this section an application of the series truncation method to the computation of the solitary wave of Fig. 3. Gravity g is taken into account but surface tension is neglected. We take a frame of reference moving with the wave, so that the flow is steady. The free surface profile is flat in the far field (since surface tension is neglected) and we introduce cartesian coordinates so that the level of the free surface in the far field corresponds to $y = 0$.

As in the previous sections, we introduce the complex potential $f = \phi + i\psi$ and the complex velocity $w = u - iv$. We choose $\phi = 0$ at the crest and $\psi = 0$ on the free surface. As $|x| \to \infty$, the flow approaches a uniform stream characterised by a constant velocity c and a constant depth h. We introduce dimensionless variables by taking c as the unit velocity and h as the unit length. On the free surface, the Bernoulli equation yields

$$
u^2 + v^2 + \frac{2}{F^2} y = 1. \tag{98}
$$

Here F is the Froude number defined by

$$F = \frac{c}{(gh)^{1/2}}. \tag{99}$$

We first map the flow domain in the region $|t| < 1$ of the complex t-plane by the transformation

$$f = \frac{2}{\pi} \ln \frac{1+t}{1-t} - i. \tag{100}$$

The transformation (100) maps the bottom of the channel onto the real diameter $-1 < t < 1$ and the free surface onto the half circumference $|t| = 1$ in the upper half t-plane. We use the notation $t = re^{i\sigma}$, so that the free surface is described by $r = 1$ and $0 < \sigma < \pi$.

Hunter and Vanden-Broeck (1983b) calculated the highest solitary wave by representing the complex velocity w by the expansion

$$w = \left[\frac{1+t^2}{2} \right]^{1/3} \exp \left[A(1-t^2)^{2\lambda} + \sum_{n=1}^{\infty} a_n(t^{2n} - 1) \right], \tag{101}$$

where λ is the smallest positive root of

$$\pi\lambda - \frac{\tan \pi\lambda}{F^2} = 0. \tag{102}$$

The term $(1 - t^2)^{2\lambda}$ in (101) results from the fact that the complex velocity on the free surface satisfies

$$u - iv \approx 1 + Ae^{-\pi\lambda|\phi|}, \quad \text{as } |\phi| \to \infty. \tag{103}$$

The constants a_n and A have to be found to satisfy the free surface condition (98).

The analysis leading to (93) shows that solitary waves of arbitrary amplitude (including the highest) can be calculated by representing w^3 by

$$w^3 = 1 + (1-t^2)^{2\lambda} \sum_{n=1}^{\infty} a_n t^{2n-2} \tag{104}$$

or by representing w by

$$w = \left[\frac{1+\beta t^2}{1+\beta} \right]^{1/3} \exp \left[A(1-t^2)^{2\lambda} + \sum_{n=1}^{\infty} a_n(t^{2n} - 1) \right]. \tag{105}$$

Here, β is another constant to be found as part of the solution. The expansion (101) is similar to the Michell expansion for the periodic wave of largest steepness (see (91) and was used before by Lenau (1966). The factor $(1 + t^2)^{1/3}$ removes the singularity associated with the 120° angle at the crest of the highest wave. The expansions (103) and (104) are equivalent to the Davies (1951) approximation and the Havelock (1919) expansion (see Vanden-Broeck and Miloh 1995).

As shown by Vanden-Broeck and Miloh (1995), equivalent numerical results are obtained by using (104) and (105). Here we describe results based on (105).

First, we differentiate (98) with respect to σ. Using (100), we obtain

$$F^2[u(\sigma)u_\sigma(\sigma) + v(\sigma)v_\sigma(\sigma)] - \frac{2}{\pi} \frac{v(\sigma)}{u^2(\sigma) + v^2(\sigma)} \frac{1}{\sin \sigma} = 0. \qquad (106)$$

Following Byatt-Smith and Longuet-Higgins (1976), we characterised the amplitude of the solitary wave by the parameter

$$\omega = 1 - F^2[(u(0)]^2. \qquad (107)$$

Here, $u(0)$ is the velocity at the rest of the wave. The parameter ω is a measure of the amplitude of the solitary wave and $\omega = 1$ for the highest wave.

We now truncate the infinite series in (104) after N terms and determine the $N + 4$ unknowns $a_1, a_2, \ldots, a_N, A, \beta, \lambda$ and F^2 by collocation. Thus we introduce the $N + 2$ collocation points

$$\sigma_I = \frac{E}{2} + (I - 1)E, \quad I = 1, 2, \ldots, N + 2, \qquad (108)$$

where $E = \pi/(2N + 4)$. We satisfy (98) at the collocation points (108). This yields $N + 2$ nonlinear algebraic equations. The last two equations are given by (102) and (107). Thus for a given value of ω, we have a system of $N + 4$ equations with $N + 4$ unknowns. This system is solved by Newton's iterations. In some of the calculations, it is more convenient to use a variant of the scheme where β is fixed and ω is found as part of the solution.

As an example, we present the numerical results for the highest wave. Then we fix $\omega = 1$ (or $\beta = 1$, if we use the variant of the scheme with β fixed). The scheme is then equivalent to that used by Hunter and Vanden-Broeck (1983b). We note that for the highest wave (for which $u = v = 0$ at the crest), the elevation α of the crest is related to F by the simple relation

$$F^2 = \frac{2}{\alpha}. \qquad (109)$$

Hunter and Vanden-Broeck (1983b) computed solutions for values of $N \le 100$ and concluded that $\alpha = 0.83322$. This estimate can be improved by increasing N. We found the values 0.83321702, 0.83320877, 0.83320544, 0.83320315 and 0.83320200 for $N = 200, 400, 600, 900$ and 1200 respectively. An extrapolation for $N \to \infty$ gives the value 0.8331986. This value is in close agreement with the value 0.833197 obtained by Williams (1981) and the value 0.83319979 obtained by Evans and Ford (1996).

7 Extensions

In the previous sections, we described two basic methods (boundary integral equation method and series truncation method) to compute periodic and solitary waves. In this section, we present various extensions of the boundary integral equation method.

All the configurations considered so far neglect the motion in the upper fluid and assume that it is characterised by a constant atmospheric pressure. The boundary integral equation method can be extended to the problem of interfacial waves propagating at the interface of two fluids of constant densities. There are then two integral relations (one for each fluid) similar to equation A. The values of $\tilde{\alpha}$ and $\tilde{\beta}$ on each side of the interface are then related by equating the difference of pressures across the interface to the product of the surface tension with the curvature of the interface. The pressures in each fluid are themselves expressed in terms of $\tilde{\alpha}$ and $\tilde{\beta}$ by using Bernoulli equations in each fluid. This approach (and variants of it) was used by many previous investigators (see, for example, Turner and Vanden-Broeck 1986, 1988; Pullin and Grimshaw 1988; Sha and Vanden-Broeck 1993; Dias and Vanden-Broeck 2004).

The results in the previous sections assume that the flows are irrotational. This is often a good assumption. However, vorticity can be generated (e.g. by wind stress on the free surface). The boundary integral equation method can be extended to flows with constant vorticity. Then the stream function ψ satisfie

$$\nabla^2 \psi = -\Omega, \tag{110}$$

where Ω is the constant vorticity. If we write

$$\psi = \psi_p + \Psi, \tag{111}$$

where ψ_p is an exact solution of (110), then Ψ satisfies Laplace's equation. We can then use the theory of analytic functions to formulate the problem in terms of Ψ (the resulting equations are similar to those in Section 5.2). In the calculations, simple choices are made for ψ_p. For example,

$$\psi_p = -\Omega \frac{y^2}{2} + Ay + B, \tag{112}$$

where A and B are constants. This idea was used by Simmen and Saffman (1985), Pullin and Grimshaw (1988), Teles da Silva and Peregrine (1988), Vanden-Broeck (1994, 1995, 1996), McCue and Forbes (2002) and Kang and Vanden-Broeck (2002).

A third recent extension is the study of waves in the presence of electric fields. The electric potentials in each fluid satisfies Laplace's equation. The values of the electric potential on each side of the interface satisfy then integral equations similar to those of Section 5.2. The coupling between hydrodynamics and the electric fields occur in the dynamic boundary condition, which involves the Maxwell stresses. This usually leads to large systems of integro-differential equations for which accurate numerical methods have been devised (Papageorgiou and Vanden-Broeck 2003, 2004).

We mentioned at the beginning of this chapter that many free surface flow problems can be approximated as two dimensional. There are of course many others which cannot (an example is the free surface flow generated by a moving ship). For three-dimensional flows, the theory of analytic functions is no longer available. However, boundary integral equation methods can be derived by using Green's theorem and Green's functions. Such methods were derived by Forbes (1989), Parau and Vanden-Broeck (2002) and others. They were generalised to include the effect of surface tension by Parau *et al.* (2005a, 2005b). The numerical computations show that there are three-dimensional gravity–capillary waves. These waves are the three-dimensional equivalent of the two-dimensional waves shown in Figs 4 and 5. They have decaying oscillations in the direction of propagation and monotonic decay in the direction perpendicular to the direction of propagation. There are both elevation and depression waves. Typical free surface profiles are shown in Figs 11 and 12. These numerical results are consistent with the asymptotic findings of Kim and Akylas (2005, 2006) and Milewski (2005) and with the rigorous analytical results of Groves and Sun (2006).

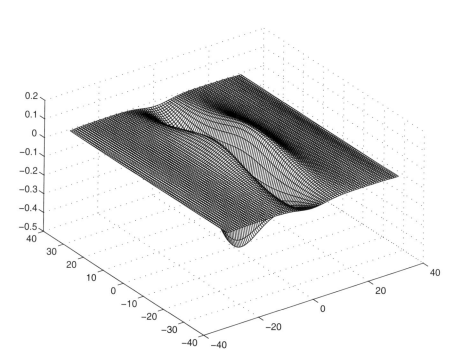

Figure 11: Computed free surface profile of a three-dimensional depression solitary wave for $gh/c^2 = 1.09$ and $T/\rho hc^2 = 0.27$.

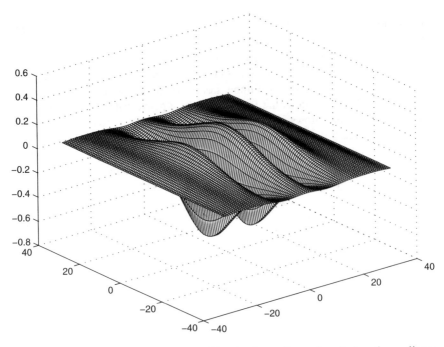

Figure 12: Computed free surface profile of a three-dimensional elevation solitary wave for $gh/c^2 = 1.13$ and $T/\rho hc^2 = 0.27$.

Finally let us mention that the results presented in this chapter were restricted to steady free surface flows. There is in addition an extensive literature on time-dependent free surface flows and stabilty. For three-dimensional time-dependent numerical calculations and references, see, for example, Grilli *et al.* (2001).

References

Akylas, T.R., Envelope solitons with stationary crests. *Phys. Fluids*, **A5**, pp. 789–791, 1993.

Akylas, T.R. & Grimshaw, R., Solitary internal waves with oscillatory tails. *J. Fluid Mech.*, **242**, pp. 279–298, 1992.

Amick, C.J., Fraenkel, L.E. & Toland, J.F., On the Stokes conjecture and the wave of extreme form. *Acta Math.*, **148**, pp. 193–214, 1982.

Beale, T.J., Solitary water waves with capillary ripples at infinity. *Comm. Pure Appl. Maths*, **64**, pp. 211–257, 1991.

Blyth, M. & Vanden-Broeck, J.-M., New solutions for capillary waves on fluid sheets. *J. Fluid Mech.*, **507**, pp. 255–264, 2004.

Byatt-Smith, J.G.B. & Longuet-Higgins, M.S., On the speed and profile of steep solitary waves. *Proc. R. Soc. Lond. A*, **350**, pp. 175–189, 1976.

Champneys, A.R., Vanden-Broeck, J.-M. & Lord, G.J., Do true elevation gravity-capillary solitary waves exist? A numerical investigation. *J. Fluid Mech.*, **454**, pp. 403–417, 2002.

Chen, B. & Saffman, P.G., Steady gravity–capillary waves on deep water. Part I: Weakly nonlinear waves. *Stud. Appl. Math.*, **60**, pp. 183–210, 1979.

Chen, B. & Saffman, P.G., Numerical evidence for the existence of new types of gravity waves on deep water. *Stud. Appl. Math.*, **62**, pp. 1–21, 1980a.

Chen, B. & Saffman, P.G., Steady gravity-capillary waves on deep water. Part II: Numerical results for finite amplitude. *Stud. Appl. Math.*, **62**, pp. 95–111, 1980b.

Cokelet, E.D., Steep gravity waves in water of arbitrary uniform depth. *Philos. Trans. Roy. Soc. London Ser. A*, **286**, pp. 183–230, 1977.

Crapper, G.D., An exact solution for progressive capillary waves of arbitrary amplitude. *J. Fluid Mech.*, **2**, pp. 532–540, 1957.

Davies, T.V., Theory of symmetrical gravity waves of finite amplitude. *Proc. Roy. Soc. London Ser. A*, **208**, pp. 475–486, 1951.

Dias, F. & Iooss, G., Capillary-gravity solitary waves with damped oscillations. *Physica D*, **65**, pp. 399–423, 1993.

Dias, F. & Kharif, C., Nonlinear gravity and capillary–gravity waves. *Ann. Rev. Fluid Mech.*, **31**, pp. 301–346, 1999.

Dias, F. & Vanden-Broeck, J.-M., Two-layer hydraulic falls over an obstacle. *European Journal of Mechanics B/Fluids*, **23**, pp. 879–898, 2004.

Dias, F., Menasce, D. & Vanden-Broeck, J.-M., Numerical study of capillary-gravity solitary waves. *Eur. J. Mech., B/Fluids*, **15**, pp. 17–36, 1996.

Evans, W.A.B. & Ford, M.J., An exact integral equation for solitary waves (with new numerical results for some 'internal' properties). *Proc. R. Soc. Lond. A*, **452**, pp. 373–390, 1996.

Forbes, L.K., An algorithm for 3-dimensional free-surface problems in hydrodynamics. *Journal of Computational Physics*, **82**, pp. 330–347, 1989.

Grilli, S.T., Guyenne, P. & Dias, F., A fully non-linear model for three-dimensional overturning waves over an arbitrary bottom. *Int. J. Numer. Meth. Fluids*, **35**, pp. 829–867, 2001.

Groves, M.D. & Sun, M.S., Fully localised solitary-wave solutions of the three-dimensional gravity-capillary water-wave problem. *Arch. Rat. Mech. Anal.*, 2006 *(submitted)*.

Havelock, T.H., Periodic irrotational waves of finite amplitude. *Proc. Roy. Soc. London Ser. A*, **95**, pp. 38–51, 1919.

Hogan, S.J., Some effects of surface tension on steep water waves. Part 2. *J. Fluid Mech.*, **96**, pp. 417–445, 1980.

Hunter, J.K. & Scherule, J., Existence of perturbed solitary wave solutions to a model equation for water waves. *Physica D*, **32**, pp. 253–268, 1988.

Hunter, J.K. & Vanden-Broeck, J.-M., Solitary and periodic gravity-capillary waves of finite amplitude. *J. Fluid Mech.*, **134**, pp. 205–219, 1983a.

Hunter, J.K. & Vanden-Broeck, J.-M., Accurate computations for steep solitary waves. *J. Fluid Mech.*, **136**, pp. 63–71, 1983b.

Iooss, G. & Kirchgassner, K., Bifurcation d'ondes solitaires en présences d'une faible tension superficielle. *CR Acad. Sci. Paris*, **311(I)**, pp. 265–268, 1990.

Iooss, G. & Kirchgassner, K., Water waves for small surface tension: an approach via normal form. *Proc. Roy. Soc. Edinburgh*, **122A**, pp. 267–299, 1992.

Kang, Y. & Vanden-Broeck, J.-M., Stern waves with vorticity. *ANZIAM J.*, **43**, pp. 321–332, 2002.

Kim. B. & Akylas, T.R., On gravity-capillary lumps. *J. Fluid Mech.*, **540**, pp. 337–351, 2005.

Kim. B. & Akylas, T.R., On gravity-capillary lumps. Part 2: Two dimensional Benjamin equation. *J. Fluid Mech.*, **557**, pp. 237–256, 2006.

Kinnersley, W., Exact large amplitude capillary waves on sheets of fluid. *J. Fluid Mech.*, **77**, pp. 229–241, 1976.

Korteweg, D.J. & De Vries, G., On the change of form of long waves advancing in a rectangular channel and on a new type of long stationary waves. *Phil. Mag.*, **39**, pp. 422–443, 1895.

Lenau, C.W., The solitary wave of maximum amplitude. *J. Fluid Mech.* **26**, pp. 309–320, 1966.

Lombardi, E., *Oscillatory Integrals and Phenomena Beyond All Algebraic Orders: With Applications to Homoclinic Orbits in Reversible Systems, Lecture Notes in Mathematics,* Vol. **1741**, Springer: New York, 2000.

Longuet–Higgins, M.S., Integral properties of periodic gravity waves of finite amplitude. *Proc. Roy. Soc. Ser. A*, **342**, pp. 157–174, 1975.

Longuet-Higgins, M.S., Capillary-gravity waves of solitary type on deep water. *J. Fluid Mech.*, **200**, pp. 451–478, 1989.

Longuet-Higgins, M.S., Capillary-gravity waves of solitary type and envelope solitons on deep water. *J. Fluid Mech.*, **252**, pp. 703–711, 1993.

Longuet-Higgins, M.S. & Fox, M.J.H., Theory of the almost-highest wave. Part 2. Matching and analytical extension. *J. Fluid Mech.*, **85**, pp. 769–786, 1978.

McCue, S.W. & Forbes, L.K., Free surface flows emerging from beneath a semi-infinite plate with constant vorticity. *J. Fluid Mech.*, **461**, pp. 387–407, 2002.

Michallet, H. & Dias, F., Numerical study of generalized interfacial solitary waves. *Phys. Fluids* **11**, pp. 1502–1511, 1999.

Michell, J.H., The highest wave in water. *Philos. Mag.*, **36**, pp. 430–437, 1883.

Milewski, P.A., Three-dimensional localized solitary gravity-capillary waves. *Comm. Math. Sc.*, **3**, pp. 89–99, 2005.

Nayfeh, A.H., Triple and quintuple-dimpled wave profiles in deep water. *J. Fluid Mech.*, **13**, pp. 545–550, 1970.

Olfe, D.B. & Rottman, J.W., Some new highest-wave solutions for deep–water waves of permanent form. *J. Fluid Mech.*, **100**, pp. 801–810, 1980.

Papageorgiou, D.T. & Vanden-Broeck, J.-M., Large amplitude capillary waves in electrified fluid sheets. *J. Fluid Mech.*, **508**, pp. 71–88, 2003.

Papageorgiou, D.T. & Vanden-Broeck, J.-M., Antisymmetric capillary waves in electrified fluid sheets. *Eur. J. Appl. Math.*, **15**, pp. 609–623, 2004.

Parau, E. & Vanden-Broeck, J.-M., Nonlinear two- and three-dimensional free surface flows due to moving disturbances. *Eur. J. Mechanics B/Fluids*, **21**, pp. 643–656, 2002.

Parau, E., Vanden-Broeck, J.-M. & Cooker, M., Nonlinear three dimensional gravity capillary solitary waves. *J. Fluid Mech.*, **536**, pp. 99–105, 2005a.

Parau, E., Vanden-Broeck, J.-M. & Cooker, M., Three-dimensional gravity-capillary solitary waves in water of finite depth and related problems. *Phys. Fluids*, **17**, pp. 122101, 2005b.

Pullin, D.I. & Grimshaw, R.H.J., Finite amplitude solitary waves at the interface between two homogeneous fluids. *Phys. Fluids*, **31**, pp. 3550–3559, 1988.

Schwartz, L.W., Computer extension and analytic continuation of Stokes' expansion for gravity waves. *J. Fluid Mech.*, **62**, pp. 553–578, 1974.

Schwartz, L.W. & Vanden-Broeck, J.-M., Numerical solution of the exact equations for capillary–gravity waves. *J. Fluid Mech.*, **95**, pp. 119–139, 1979.

Sha, H. & Vanden-Broeck, J.-M., Two-layer flows past a semicircular obstacle. *Phys. Fluids A*, **5**, pp. 2661–2668, 1993.

Sha, H. & Vanden-Broeck, J.-M., Internal solitary waves with stratification in density. *J. Austr. Math. Soc. B*, **38**, pp. 563–580, 1997.

Simmen, J.A. & Saffman, P.G., Steady deep water waves on a linear shear current. *Stud. Appl. Maths*, **75**, pp. 35–57, 1985.

Stokes, G.G., On the theory of oscillatory waves. *Camb. Trans.*, **8**, pp. 441–473, 1847.

Sun, S.M., Existence of generalized solitary wave solution for water with positive Bond number less than 1/3. *J. Math. Anal. Appl.*, **156**, pp. 471–504, 1991.

Sun, S.M., Nonexistence of truly solitary waves in water with small surface tension. *Proc. R. Soc. Lond. A*, **455**, pp. 2191–2228, 1999.

Sun, S.M. & Shen, M.C., Exponentially small estimate for the amplitude of capillary ripples of generalised solitary waves. *J. Math. Anal. Appl.*, **172**, pp. 533–566, 1993.

Teles da Silva, A.F. & Peregrine, D.H., Steep solitary waves in water of finite depth with constant vorticity. *J. Fluid Mech.*, **195**, pp. 281–305, 1988.

Turner, R.E.L. & Vanden-Broeck, J.-M., The limiting configuration of interfacial gravity waves. *Phys. Fluids*, **29**, pp. 372–375, 1986.

Turner, R.E.L. & Vanden-Broeck, J.-M., Broadening on interfacial solitary waves. *Phys. Fluids*, **31**, pp. 2486–2490, 1988.

Turner, R.E.L. & Vanden-Broeck, J.-M., Long internal waves. *Phys. Fluids*, **A4**, pp. 1929–1935, 1992.

Vanden-Broeck, J.-M., Some new gravity waves in water of finite depth. *Phys. Fluids*, **26**, pp. 2385–2387, 1983.

Vanden-Broeck, J.-M., Steep gravity waves: Havelock's method revisited. *Phys. Fluids*, **29**, pp. 3084–3085, 1986.

Vanden-Broeck, J.-M., Elevation solitary waves with surface tension. *Phys. Fluids*, **A3**, pp. 2659–2663, 1991.

Vanden-Broeck, J.-M., Steep solitary waves in water of finite depth with constant vorticity. *J. Fluid Mech.*, **274**, pp. 339–348, 1994.

Vanden-Broeck, J.-M., New families of steep solitary waves in water of finite depth with constant vorticity. *Eur. J. Mech. B/fluids*, **14**. pp. 761–774, 1995.

Vanden-Broeck, J.-M., Periodic waves with constant vorticity in water of infinite depth. *IMA J. Appl. Math.* **56**, pp. 207–217, 1996.

Vanden-Broeck, J.-M. & Dias, F., Gravity-capillary solitary waves in water of infinite depth and related free-surface flows. *J. Fluid Mech.*, **240**, 549–557, 1992.

Vanden-Broeck, J.-M. & Keller, J.B., A new family of capillary waves. *J. Fluid Mech.*, **98**, pp. 161–169, 1980.

Vanden-Broeck, J.-M. & Miloh, T., Computations of steep gravity waves by a refinement of the Davies-Tulin's approximation. *SIAM's J. Appl. Math.*, **55**, pp. 892–903, 1995.

Vanden-Broeck, J.-M. & Schwartz, L.W., Numerical computation of steep gravity waves in shallow water. *Phys. Fluids*, **22**, pp. 1868–1871, 1979.

Williams, J.M., Limiting gravity waves in water of finite depth. *Phil. Trans. R. Soc. Lond. A*, **302**, pp. 139–188, 1981.

Wilton, J.R., On ripples. *Phil. Mag.*, **29**, pp. 688–700, 1915.

Zufuria, J.A., Symmetry breaking in periodic and solitary gravity-capillary waves on water of finite depth. *J. Fluid Mech.*, **184**, pp. 183–206, 1987.

CHAPTER 4

Internal solitary waves

E. Pelinovsky, O. Polukhina, A. Slunyaev & T. Talipova
*Institute of Applied Physics, Russian Academy of Sciences,
Nizhny Novgorod, Russia.*

Abstract

In this chapter we review various approximate models (from weakly to strongly non-linear) describing internal solitary waves in the density-stratified fluids. The waves may have different shapes and polarities depending on the fluid stratification. These properties as well as the processes of generation, interaction and transformation are discussed.

1 Introduction

The physical nature of waves propagating on the boundary between two layers of fluids of different densities is well known and explained by particle motion under the action of gravity and inertial forces; the familiar example of such waves is the surface waves (e.g. wind waves, swells, tides, tsunamis and so on) seen in seas and oceans. However, the density and shear flow in the ocean and atmosphere are stratified in the vertical direction, and hence *internal* waves can propagate at various depths in the ocean and at various heights in the atmosphere. Due to the rather weak density variation, the restoring force is also weak, and very large vertical displacements of fluid particles may be achieved by the impact of external forces. The amplitudes of natural internal waves can reach 100 m, significantly exceeding the limiting amplitudes of the waves on the sea–air boundary (such as wind waves and swells). Figure 1 gives an example of an observation of large-amplitude (up to 40 m) long internal waves in the North Atlantic to the north of Ireland (Small *et al.* 1999). Many intense waves of this kind have solitary-like shapes that propagate for a long time without significant change of energy. They can be interpreted and well described as internal *solitary waves* (*solitons*). Strictly speaking, the term

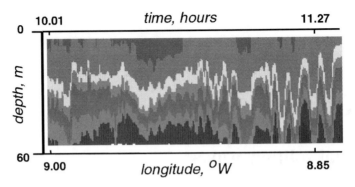

Figure 1: A group of intense internal waves observed in the Atlantic, near Ireland (Small *et al.* 1999).

'soliton' is reserved for solitary waves in integrable systems, but we will neverthe-less follow the widely used custom and call these waves internal solitons. As we will describe, such a terminology is partially justified here because these waves can be successfully modelled by integrable equations of the Korteweg–de Vries (KdV) type. Observations of large-amplitude long internal waves in the environment are compiled in several reviews: in the oceans (Ostrovsky and Stepananyats 1989; Jeans 1995; Holloway *et al.* 2002; Global Ocean Associates 2004; Sabinin and Serebranny 2005; Helfrich and Melville 2006), in the earth's atmosphere (Cheung and Little 1990; Rottman and Grimshaw 2002) and in stratified laboratory tanks (Ostrovsky and Stepanyants 2005). Various appropriate mathematical models have been devel-oped to describe nonlinear internal waves (Grimshaw 2002). The first study of inter-nal solitons was based on the nonlinear boundary problem for the stream function (Dubriel-Jacotin 1932), and this approach remains popular in numerical simulations of large-amplitude solitons. The unsteady dynamics of small-amplitude soliton-like internal waves was first analysed in the framework of the KdV equation (Benney 1966), then more complicated models based on the extended versions of the KdV and Boussinesq equations have been developed for waves of moderate and large amplitudes. Now, the direct numerical simulation of the basic 2D Euler equations is being actively carried out (Grue *et al.* 1999; Lamb 2002; Vlasenko *et al.* 2005) to study the generation, propagation and breaking of internal solitons.

This chapter reviews the modern approaches used to describe internal solitons. Different mathematical models (from weakly to strongly nonlinear) are summarized in Section 2. The steady-state solitary wave solutions are discussed in Section 3. It is shown that their shape critically depends on the sign of the cubic nonlinearity. Another type of localized nonlinear waves (breathers) is briefly discussed. The process of internal soliton generation from the long-scale disturbances is described in Section 4 within the framework of the extended KdV (Gardner) equation. The transformation of a soliton passing a transition zone, where the variable stratification leads to the change of signs of the model coefficients (which often happens on ocean shelves) is discussed in Section 5.

2 Approximate evolution equations

Weakly nonlinear models for long internal waves are based on the KdV equation and its generalizations (Benney 1966; Lee and Beardsley 1974; Kakutani and Yamasaki 1978; Lamb and Yan 1996; Grimshaw 2002; Holloway *et al.* 2002 and Grimshaw *et al.* 2002b). Such equations are derived by using a multiscale asymptotic procedure on the governing Euler equations for inviscid incompressible stratified fluids. For simplicity, we consider a two-dimensional flow; then the basic equations are:

$$\rho\frac{du}{dt} + \frac{\partial P}{\partial x} = 0, \quad \rho\frac{dw}{dt} + \frac{\partial P}{\partial z} + \rho g = 0, \tag{1}$$

$$\frac{d\rho}{dt} = 0, \quad \frac{\partial u}{\partial x} + \frac{\partial w}{\partial z} = 0, \tag{2}$$

where $\vec{V} = \{u, w\}$ is the fluid velocity, $\rho(x, z, t)$ is the fluid density, P is the pressure, g is gravitational acceleration, $\{x, z\}$ are spatial coordinates (horizontal and vertical – see Fig. 2) and d/dt is the convective time derivative. The fluid is assumed to be confined between two rigid boundaries: $z = -h$ (bottom) and $z = 0$ (free surface). In fact, the approximation of the rigid boundary on the free surface 'works' very well in natural water basins due to the weak variation of the water density compared with the density jump on the surface. Somewhat more laborious computations are necessary to take into account the free moving surface (Grimshaw *et al.* 2002b). Thus, the boundary conditions for equations (1) and (2) have the form:

$$w = 0 \text{ at } z = -h \quad \text{and} \quad z = 0. \tag{3}$$

Since the density of the fluid particles does not change, it is convenient to introduce the isopycnal (Lagrangian) coordinate

$$y = z - \zeta(x, z, t), \tag{4}$$

where $\zeta(x, z, t)$ is the vertical displacement of a fluid particle from its rest position. Thus, density $\rho(x, z, t) = \rho_0(y)$ is 'frozen' in this representation. We also assume

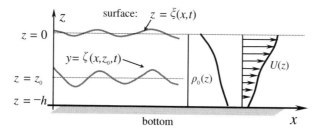

Figure 2: Coordinate system and the background flow ($\xi = 0$ in the rigid boundary approximation).

that the horizontal velocity field $u(x, y, t)$ can be decomposed into the unperturbed horizontal shear flow velocity $U(y)$ and its perturbation $u'(x, y, t)$: $u = U + u'$.

Let us consider the weakly nonlinear limit, when the wave amplitude is small with respect to depth (with the corresponding small parameter $\mu \ll 1$ but finite. Another small parameter $\varepsilon = h/L$ (L is the characteristic wavelength scale and h is the total depth) characterizes the weak dispersion (long waves) and defines slow variables: $X = \varepsilon x$, $T = \varepsilon t$. We anticipate the KdV-scaling $\mu = \varepsilon^2$ to implement the contribution of the first nonlinear and dispersive corrections to the wave equation in the same order of smallness. To enable building a higher-order evolution equation, we introduce also the set of new temporal variables $\tau_i = \mu^i T$:

$$\frac{\partial}{\partial T} = -c\frac{\partial}{\partial s} + \mu\frac{\partial}{\partial \tau}, \quad \frac{\partial}{\partial \tau} = \frac{\partial}{\partial \tau_1} + \mu\frac{\partial}{\partial \tau_2} + \cdots,$$

$$\frac{\partial}{\partial X} = \frac{\partial}{\partial s}, \quad s = X - cT.$$

(5)

The co-moving coordinate s contains the linear long-wave speed c, which is yet to be determined.

After tedious, but straightforward, calculations the systems (1) and (2) lead to one equation containing ζ only,

$$\frac{\partial}{\partial y}\left\{\rho_0(c - U)^2\frac{\partial^2\zeta}{\partial s\partial y}\right\} + \rho_0 N^2\frac{\partial\zeta}{\partial s} = M,$$

(6)

where $N(y)$ is the Brunt–Väisälä (buoyancy) frequency profile, $N^2(y) = -(g/\rho_0(y))(d\rho_0/dy)$, and M is a complicated expression containing nonlinear and dispersive terms of different orders (Grimshaw et al. 2002b). The boundary conditions for equation (6) are $\zeta = 0$, at the bottom $y = -h$ and on the surface $y = 0$.

Next, we assume that the isopycnal displacement ζ is represented by the asymptotic series

$$\zeta(s, y, \tau) = \mu A(s, \tau)\Phi(y) + \mu^2\zeta_2(s, y, \tau) + \mu^3\zeta_3(s, y, \tau) + \cdots,$$

(7)

where all the functions should be determined after substituting the series (7) into equation (6) and collecting the terms of the same order of μ. In the lowest order, the linear eigenvalue problem for the modal function $\Phi(y)$ is obtained:

$$L\Phi \equiv \frac{d}{dy}\left[\rho_0(c - U)^2\frac{d\Phi}{dy}\right] + \rho_0 N^2(y)\Phi = 0,$$

(8)

with zero boundary conditions on the bottom and the surface of the fluid. The solution of (8) may be simplified by neglecting the weak density variation in natural water and taking $\rho_0(y) \equiv \rho_0$ a constant; this corresponds to the Boussinesq approximation, which will be used hereafter. Real solutions (stable internal waves) of equation (8) can be obtained only for large values of the Richardson number ($Ri = N^2/U_z^2 > 1/4$) (see, for instance, Miles 1961). It is well known that the eigenvalue problem (8) has, in general, an infinite sequence of modes $\Phi_n^{\pm}$ with corresponding speeds (eigenvalues) $c_n^{\pm}$ ($n = 0, 1, 2, \ldots$); here the modes with $c_n^{\pm}$ propagate faster/slower than the background current. A formal theory can be developed for any of these modes, but the amplitude of an internal soliton belonging to a higher mode gradually decreases due to the energy transfer to radiated shorter lower-mode dispersive waves (Akylas and Grimshaw 1992). This is why theories are developed mainly for solitary waves of the lowest ($n = 0$) mode only, which has the greatest speed c. The corresponding modal function $\Phi(y)$ is defined by (8). We choose the normalization of the modal function in a way that its extreme value is unity, i.e. $\Phi_{\max} = 1$ at $y_{\max}$.

The compatibility conditions to solve each order of the inhomogeneous problem (6) give the sequence of evolution equations defining the amplitude dynamics up to different orders of accuracy. At $O(\mu^2)$, we obtain the famous KdV equation derived first in this context by Benney (1966):

$$\frac{\partial A}{\partial \tau} + \alpha A \frac{\partial A}{\partial s} + \beta \frac{\partial^3 A}{\partial s^3} = 0, \tag{9}$$

where

$$\alpha = \frac{3}{2I} \int (c - U)^2 (d\Phi/dy)^3 dy, \quad \beta = \frac{1}{2I} \int (c - U)^2 \Phi^2 dy,$$

$$I = \int (c - U)(d\Phi/dy)^2 dy, \tag{10}$$

and all integrals are calculated over the fluid depth. The KdV equation is a very popular model to demonstrate the soliton properties of the internal wave field in the ocean and atmosphere.

After solving the compatibility condition for equation (6) in order $O(\mu^2)$ with respect to ζ_2, one obtains

$$\zeta_2 = A^2 T_n(y) + \frac{\partial^2 A}{\partial s^2} T_d(y), \tag{11}$$

where $T_n(y)$ is the first nonlinear and $T_d(y)$ is the first dispersion correction to the modal structure $\Phi(y)$ of the internal wave; these are solutions of corresponding inhomogeneous boundary problems:

$$L T_n = -\alpha \frac{d}{dy} \left\{ \rho_0 (c - U) \frac{d\Phi}{dy} \right\} + \frac{3}{2} \frac{d}{dy} \left\{ \rho_0 (c - U)^2 \left(\frac{d\Phi}{dy} \right)^2 \right\} \tag{12}$$

and

$$L T_d = -2\beta \frac{d}{dy} \left\{ \rho_0 (c - U) \frac{d\Phi}{dy} \right\} - \rho_0 (c - U)^2 \Phi, \tag{13}$$

with zero boundary conditions on the bottom and the surface of the fluid. It is important to note that solutions of the boundary-value problems (12) and (13) are unique only up to additive multiples of Φ. It is convenient to let $A(s, \tau)$ represent the isopycnal displacement at the level $y_{\max}$. Hence, we choose the auxiliary conditions: $T_n(y_{\max}) = T_d(y_{\max}) = 0$. In this case, series (7), at the level $y = y_{\max}$ gives

$$\zeta(s, y_{\max}, \tau) = \mu A(s, \tau) + O(\mu^3); \tag{14}$$

of course, other normalizations can be used if convenient.

The compatibility condition for (6) at $O(\mu^3)$, along with (5) and (9), leads to the second-order KdV equation:

$$\frac{\partial A}{\partial \tau} + \alpha A \frac{\partial A}{\partial s} + \beta \frac{\partial^3 A}{\partial s^3} + \mu \left(\alpha_1 A^2 \frac{\partial A}{\partial s} + \beta_1 \frac{\partial^5 A}{\partial s^5} + \gamma_1 A \frac{\partial^3 A}{\partial s^3} + \gamma_2 \frac{\partial A}{\partial s} \frac{\partial^2 A}{\partial s^2} \right) = 0, \tag{15}$$

where the new coefficients are given by

$$\alpha_1 = \frac{1}{2I} \int dy \left\{ 3(c - U)^2 \left[3(dT_n/dy) - 2(d\Phi/dy)^2 \right] (d\Phi/dy)^2 \right. $$
$$\left. + \alpha(c - U) \left[5(d\Phi/dy)^2 - 4(dT_n/dy) \right] (d\Phi/dy) - \alpha^2 (d\Phi/dy)^2 \right\}, \tag{16a}$$

$$\beta_1 = \frac{1}{2I} \int dy \left\{ 2\beta(c - U) \left[\Phi^2 - (d\Phi/dy)(dT_d/dy) \right] \right. $$
$$\left. - \beta^2 (d\Phi/dy)^2 + (c - U)^2 \Phi T_d \right\}, \tag{16b}$$

$$\gamma_1 = -\frac{1}{2I} \int dy \left\{ 2(c - U) \left[\alpha(dT_d/dy) + 2\beta(dT_n/dy) \right] (d\Phi/dy) \right. $$
$$+ 2\alpha\beta(d\Phi/dy)^2 - 2\alpha(c - U)\Phi^2 + (c - U)^2 \Phi^2 (d\Phi/dy) $$
$$\left. - 4\beta(c - U)(d\Phi/dy)^3 - (c - U)^2 \left[3(dT_d/dy)(d\Phi/dy)^2 + 2T_n\Phi \right] \right\}, \tag{16c}$$

$$\gamma_2 = \frac{1}{2I} \int dy \left\{ (c - U) \left[2\beta(d\Phi/dy)^3 + 6\alpha\Phi^2 \right] - 3\alpha\beta(d\Phi/dy)^2 \right. $$
$$- 2(c - U)^2 \left[\Phi^2(d\Phi/dy) - 3T_n\Phi \right] $$
$$\left. - 6\alpha(c - U)(dT_d/dy)(d\Phi/dy) + 3(c - U)^2 dT_d/dy)(d\Phi/dy)^2 \right\}. \tag{16d}$$

This asymptotic procedure allows us to obtain extensions of the KdV equation to any order. The asymptotically obtained evolution equations may acquire some specific features that are absent in the exact solutions of the Euler equations. For instance, if the linear wave solutions of the extended KdV equation (15) are considered, their wave frequency ω and wavenumber k satisfy the dispersion relation

$$\omega = ck - \mu\beta k^3 + \mu^2 \beta_1 k^5. \tag{17}$$

Since $\beta > 0$ (see (10)), a positive coefficient β_1 (at least for the case of a two-layer fluid) implies that the group velocity of short waves may exceed the speed c of long waves, in contradiction with the exact linear dispersion relation of the original physical problem. Thus, asymptotically derived equations may lead to different (wrong) wave dynamics. This takes place when the formal ranges of applicability of asymptotic expansions are not satisfied and happens due to the *truncation* of the asymptotic series. An asymptotically derived evolution equation may be written in many asymptotically close forms. All of them are equally justified, but may show different properties when the applicability conditions are broken. This freedom may be used to obtain improved models that are more preferable from this or that point of view. One regular way of obtaining the models with a more accurate dispersion law is based on the Pade approximation, which requires a modification of the asymptotic procedure. As a result, the recently derived Boussinesq-like models can be applied to fully nonlinear and almost fully dispersive surface waves (Madsen *et al.* 2003). The same approach may be applied in the case of nonlinear internal waves as well.

The extended KdV equation (15) may be reduced to the KdV equation (9) using the smallness of nonlinear and dispersive parameter μ. The following near-identity asymptotical transformation of the wave field,

$$B = A + \mu \left\{ \frac{1}{2}\lambda_1 A^2 + \lambda_2 A_{ss} + \lambda_3 A_s \int_{s_0}^{s} A \, ds + \lambda_4 XA_\tau \right\}, \qquad (18)$$

(where $\lambda_1, \ldots, \lambda_4$ are constants) can reduce the second-order evolution equation (15) to its integrable version, or to the first-order KdV equation (9) for the field B with accuracy up to $O(\mu^2)$ (see, for example, Kodama 1985; Fokas *et al.* 1996), where the transformation (18) is introduced for the particular case of surface waves. In the general case, the coefficients $\lambda_1, \ldots \lambda_4$ are given by

$$\lambda_1 = \frac{-18\beta^2\alpha_1 + 2\alpha^2\beta_1 + 3\alpha\beta\gamma_1}{9\alpha\beta^2}, \quad \lambda_2 = \frac{-6\beta^2\alpha_1 - \alpha^2\beta_1 + \alpha\beta\gamma_2}{2\alpha^2\beta},$$

$$\lambda_3 = \frac{4\alpha\beta_1 - 3\beta\gamma_1}{9\beta^2}, \quad \lambda_4 = -\frac{\beta_1}{3\beta^2}, \qquad (19)$$

assuming that nonlinear and dispersive coefficients α and β in the KdV equation (9) differ from zero. Thus, in general, equation (15) is asymptotically reducible to the integrable KdV equation (9).

The coefficients α and β of the KdV equation are determined by integral expressions (10) and can vary over a rather wide range. The dispersion coefficient β is always positive since the integrands in (19) are positive expressions. At the same time, the nonlinear coefficient α can have either sign and may even be zero. A two-layer fluid is a popular example of such a case; then $\alpha = 3c(h_1 - h_2)/(2h_1h_2)$, where h_1 and h_2 are the thicknesses of the upper and lower layers (Kakutani and Yamasaki 1978). If α approaches zero, the classical hierarchy of small parameters gets broken and the asymptotic procedure must be modified. This anomalous smallness of the nonlinear term in the KdV equation requires taking into account the next nonlinear term with the coefficient α_1 in the second-order equation (15) for the description of nonlinear processes. The other second-order terms in (15) turn out to be of the next order of smallness in this degenerate case. Such an extension of the

KdV equation is often called the *Gardner* equation. To derive the Gardner equation using the asymptotic expansions, one should explicitly take into consideration the smallness of $\alpha \sim \delta$, where $\delta \ll 1$, assuming $\delta \sim \mu$, modifying the KdV-scaling as $\mu = \varepsilon$ and revising all the series. Then the following equation results:

$$\frac{\partial A}{\partial \tau} + \alpha A \frac{\partial A}{\partial s} + \beta \frac{\partial^3 A}{\partial s^3} + \alpha_1 A^2 \frac{\partial A}{\partial s} = 0. \tag{20}$$

It is necessary to note that this equation implies a less strict condition for the smallness of wave amplitude: classical KdV assumption $\mu \ll 1$ transforms to $\mu^2 \ll 1$ for the Gardner equation (note that now $\tau \sim \mu^2 T$), and therefore, the wave amplitude can be moderate, not just weak.

Equation (20) demonstrates quite different strongly nonlinear wave dynamics determined by the sign of the cubic nonlinear term α_1, which can be negative as well as positive depending on the stratification (Grimshaw 2002). In the particular case of a two-layer fluid, it is always negative (Kakutani and Yamasaki 1978). The Gardner equation has been the subject of many investigations (Miles 1979, 1981; Grimshaw *et al.* 1999; Slyunyaev and Pelinovskii 1999; Slyunyaev 2001; Grimshaw *et al.* 2002a; Nakoulima *et al.* 2004). This equation is now widely used for modelling of internal solitons in the ocean (Liu *et al.* 1998; Small 1999; Holloway *et al.* 2002).

The extension of the Gardner equation (20) to the next order in asymptotic theory can also be derived:

$$\frac{\partial A}{\partial \tau} + \alpha A \frac{\partial A}{\partial s} + \beta \frac{\partial^3 A}{\partial s^3} + \alpha_1 A^2 \frac{\partial A}{\partial s}$$
$$+ \mu^2 \left(\alpha_2 A^3 \frac{\partial A}{\partial s} + \alpha_3 A^4 \frac{\partial A}{\partial s} + \gamma_1 A \frac{\partial^3 A}{\partial s^3} + \gamma_2 \frac{\partial A}{\partial s} \frac{\partial^2 A}{\partial s^2} \right.$$
$$\left. + \gamma_{31} \left(\frac{\partial A}{\partial s} \right)^3 + \gamma_{32} A \frac{\partial A}{\partial s} \frac{\partial^2 A}{\partial s^2} + \gamma_{33} A^2 \frac{\partial^3 A}{\partial s^3} + \beta_1 \frac{\partial^5 A}{\partial s^5} \right) = 0, \tag{21}$$

and the asymptotic transformation similar to (18) can be suggested:

$$B = A + \mu^2 \left[\lambda_1 A^2 + \lambda_2 A^3 + \lambda_3 A_{ss} + \lambda_4 A_s \int_{s_0}^s A ds + \lambda_5 A_s \int_{s_0}^s A^2 ds + \lambda_6 X A_\tau \right],$$

$$\lambda_1 = \frac{3\alpha\alpha_1\beta_1 + \alpha_1\beta(\gamma_1 - \gamma_2) + \alpha\beta(3\gamma_{31} - \gamma_{32} - \gamma_{33})}{6\alpha_1\beta^2},$$

$$\lambda_2 = \frac{2\alpha_1\beta_1 + \beta(2\gamma_{31} - \gamma_{32})}{6\beta^2}, \quad \lambda_3 = \frac{4\alpha_1\beta_1 + 3\beta(3\gamma_{31} - \gamma_{32} - \gamma_{33})}{6\alpha_1\beta},$$

$$\lambda_4 = \frac{4\alpha\beta_1 - 3\beta\gamma_1}{9\beta^2}, \quad \lambda_5 = \frac{4\alpha_1\beta_1 - 3\beta\gamma_{33}}{9\beta^2}, \quad \lambda_6 = -\frac{\beta_1}{3\beta^2}, \tag{22}$$

which gives us the equation for B in a more convenient form than (21) but still different from the Gardner equation (18) (see Slunyaev et al. 2003):

$$\frac{\partial B}{\partial \tau} + \left(\alpha B + \alpha_1' B^2 + \mu \alpha_2' B^3 + \mu \alpha_3' B^4\right)\frac{\partial B}{\partial s} + \beta \frac{\partial^3 B}{\partial s^3} = 0,$$

$$\alpha_1' = \alpha_1 + \mu\rho, \quad \rho = \frac{\alpha^2}{6\alpha_1\beta}(3\gamma_{31} - \gamma_{32} - \gamma_{33}) - \frac{\alpha\gamma_2}{6\beta} + \frac{7\alpha^2\beta_1}{18\beta^2},$$

$$\alpha_2' = \alpha_2 + \frac{2\alpha_1(\gamma_1 - 3\gamma_2) + \alpha(24\gamma_{31} - 9\gamma_{32} - 8\gamma_{33})}{18\beta} + \frac{10\alpha\alpha_1\beta_1}{9\beta^2},$$

$$\alpha_3' = \alpha_3 + \frac{\alpha_1(4\gamma_{31} - 2\gamma_{32} - \gamma_{33})}{6\beta} + \frac{5\alpha_1^2\beta_1}{9\beta^2}. \tag{23}$$

Equation (23) contains two additional nonlinear terms and conserves at least two first integrals, but has the same order of asymptotic accuracy, and is therefore preferable for numerical modelling rather than (21).

To derive fully nonlinear and dispersive generalizations of the KdV equation, the heuristic or variational approach may be applied. The nonlinear and dispersive terms with the required accuracy may be incorporated independently, as it was done for surface waves (Whitham 1974). The full dispersion term can be found at least for simplified stratification, but, generally speaking, it has an integral form (via inverse Fourier transformation). Consideration of the full nonlinearity of the wave processes leads to the nonlinear interaction of waves with different modal structures, and the number of the "coupled" evolution equations may be high. This is why a two-layer model of the density stratification having one mode only is popular in the analytical and numerical study of large-amplitude internal waves. The fully nonlinear part of the evolution equations can exactly be found in the hydrostatic (shallow-water) approximation of the two-layer model (Baines 1995; Slunyaev et al. 2003):

$$\frac{\partial \eta}{\partial t} + V_{nl}\frac{\partial \eta}{\partial x} = 0,$$

$$V_{nl}(\eta) = \sqrt{g'\frac{h_1 h_2}{h}} + 3\sqrt{\frac{g'}{h}}\left[\sqrt{\tilde{h}_1 \tilde{h}_2}\,(h_1 - h_2) - \sqrt{h_1 h_2}\left(\tilde{h}_1 - \tilde{h}_2\right)\right]\frac{\tilde{h}_1 - \tilde{h}_2}{h^2},$$

$$\tilde{h}_1 = h_1 + \eta, \quad \tilde{h}_2 = h_2 - \eta, \tag{24}$$

where indexes 1 and 2 numerate the upper and lower layers correspondingly, $g' = g\Delta\rho/\rho$ is the reduced gravity acceleration, $\Delta\rho = \rho_2 - \rho_1$ is the density jump which is small (the Boussinesq approximation), ρ_j is the average density of the layers, $h = h_1 + h_2$ is the total depth and η is the interfacial displacement. Together with the pure nonlinear and pure dispersive terms, mixed terms of nonlinear dispersion (see (15) or (21)) must be taken into consideration to improve the model. Ostrovsky and Grue (2003) suggested a heuristic form of the nonlinear dispersion, which is supposed to take into consideration the strongly nonlinear effects of long waves.

Their 'β-model' is represented by a KdV-like equation for fully nonlinear and weakly dispersive interfacial waves, expressed in the original unscaled variables,

$$\frac{\partial \eta}{\partial t} + V_{nl}\frac{\partial \eta}{\partial x} + \frac{\partial}{\partial x}\left(\beta(\eta)\frac{\partial^2 \eta}{\partial x^2}\right) = 0, \quad \beta(\eta) = \frac{1}{6}V_{nl}(\eta)(h_1 + \eta)(h_2 - \eta).$$

$$(25)$$

The nonlinear evolution equations described above are derived for unidirectional waves and are valid for media where the influence of lateral boundaries may be neglected and there are no counter-propagating waves. This is a usual approach for the consideration of oceanic and atmospheric internal waves. However, for the description of internal waves propagating in closed basins, such as lakes, bays and laboratory wave flumes, the reflection from the horizontal boundaries should be taken into account. Another example is the process of generation by external factors when the generated waves propagate in both directions from the source area. Then, two-directional (Boussinesq-type) models for internal waves can be derived. If the nonlinearity and dispersion are weak, these equations can be obtained from the hydrodynamic equations using the Galerkin's procedure for any density and flow stratification (Ostrovsky 1978; Engelbrecht et al. 1988). Basic modes in this approach are the linear non-dispersive modes, and the Boussinesq equations are:

$$\frac{\partial A}{\partial t} + \frac{\partial}{\partial x}\left[\left(h + \mu\frac{N}{2}A\right)u\right] = 0,$$

$$(26)$$

$$\frac{\partial u}{\partial t} + \frac{c^2}{h}\frac{\partial A}{\partial x} + \mu N\left[u\frac{\partial u}{\partial x} - \frac{1}{2h}\frac{\partial}{\partial t}(Au)\right] + \mu Dh\frac{\partial^3 A}{\partial x \partial t^2} = 0,$$

where $A(x, t)$ and $u(x, t)$ describe the isopycnal displacement and horizontal velocity, respectively; and $N = \alpha h/c$ and $D = \beta/ch^2$ are dimensionless parameters of nonlinearity and dispersion. The basic assumptions, as well as the vertical structure and boundary conditions, are absolutely similar to those introduced for the derivation of the KdV equation. Let us mention that the horizontal velocity field in a two-dimensional spatial domain is described to the leading order μ by the expression:

$$u'(x, y, t) = \mu cu(x, t)\frac{d\Phi}{dy}.$$

$$(27)$$

The system (26) can be transformed into more convenient forms similar to the Peregrine equations for surface waves (Peregrine 1967). If we consider unidirectional waves, the KdV equation (9) can be readily derived from (26).

For a two-layer fluid, strongly nonlinear Boussinesq-like equations can be derived in the same way as for surface waves (Miyata 1988; Choi and Camassa 1999; Ostrovsky and Grue 2003; Craig et al. 2004). The main idea here is that the wave field in each layer satisfies the linear Laplace's equation for potential flow. As a result, we may use two boundary conditions (kinematic and dynamic) on the interface to derive the nonlinear equations for interface displacement and the difference between the particle velocities above and below the interface. Next, the pressure (or potential) should be presented in the Taylor's series in the layer thickness (the small parameter here is the ratio of water depth to wavelength – the dispersion parameter). The variational approach through the Hamiltonian or Lagrangian formulation is useful here to simplify the calculations. The derived Boussinesq-like equations are now called the Choi–Camassa equations (Choi and Camassa 1999):

$$\frac{\partial(h_1 - \eta)}{\partial t} + \frac{\partial}{\partial x}[(h_1 - \eta)u_1] = 0, \qquad \frac{\partial(h_2 + \eta)}{\partial t} + \frac{\partial}{\partial x}[(h_2 + \eta)u_2] = 0, \qquad (28)$$

$$\frac{\partial(u_1 - u_2)}{\partial t} + u_1 \frac{\partial u_1}{\partial x} - u_2 \frac{\partial u_2}{\partial x} - g' \frac{\partial \eta}{\partial x} = D, \qquad (29)$$

$$D = \frac{1}{3(h_1 - \eta)} \frac{\partial}{\partial x}\left[(h_1 - \eta)^3 \left(\frac{\partial^2 u_1}{\partial x \partial t} + u_1 \frac{\partial^2 u_1}{\partial x^2} - \left\{ \frac{\partial u_1}{\partial x} \right\}^2 \right) \right]$$

$$- \frac{1}{3(h_2 + \eta)} \frac{\partial}{\partial x}\left[(h_2 + \eta)^3 \left(\frac{\partial^2 u_2}{\partial x \partial t} + u_2 \frac{\partial^2 u_2}{\partial x^2} - \left\{ \frac{\partial u_2}{\partial x} \right\}^2 \right) \right], \qquad (30)$$

where u_1 and u_2 are the fluid velocities in the upper and lower layers of thickness h_1 and h_2, respectively, g' is the reduced gravity acceleration (we use the Boussinesq approximation for fluids with almost the same densities) and η is the interface displacement. If $D = 0$, the system in (28) and (29) corresponds to the nonlinear shallow-water theory for a two-layer fluid, which remains valid for large-amplitude, but smooth, waves. The term D includes weak as well as nonlinear dispersion. Due to its smallness, some terms in expression (2) can be replaced through the non-dispersive system ($D = 0$), which is why the equations of the nonlinear-dispersive theory may look different in papers by different authors. In fact, the system (28)–(30) is similar to the Green–Naghdi system for surface waves (Green and Naghdi 1976). As has been mentioned, this approach in the case of surface waves is extended now to fully nonlinear and almost fully dispersive waves (Madsen *et al.* 2003). A similar approach may be possible for waves in a two-layer fluid, but not available yet.

3 Solitons and their properties

Solitons are famous examples of the manifestation of nonlinear effects in internal wave dynamics. Discovered first for surface waves at the beginning of the 19th century, later they were found in many other physical problems (see Chapter 1). These surface solitary waves are described by the KdV and Boussinesq models, which when derived for internal waves became the first practical approximations for the observed internal solitary waves. Let us begin our consideration of solitons from the Gardner equation, bearing in mind that the KdV equation is a particular case of the Gardner equation (20) (its low-amplitude limit). The Gardner equation contains the cubic nonlinear term (α_1), and the existence of solitons depends drastically on its sign. The solitary wave for the Gardner equation can be written in a most general form as

$$A(s,\tau) = \frac{6\gamma^2/\alpha}{1 + R\cosh(\gamma(s - s_0 - V\tau))}, \quad R^2 = 1 + \frac{6\alpha_1\beta\gamma^2}{\alpha^2}, \quad V = \beta\gamma^2, \quad (31)$$

where the wave amplitude is

$$a = \frac{6\gamma^2/\alpha}{1 + R}. \quad (32)$$

The solitary wave amplitude a (or its scale coefficient γ) is a free parameter; another free parameter is the initial position of the wave s_0. It is convenient for our purposes to employ the dimensionless version of the Gardner equation (GE). After the changes (assuming that $\alpha \neq 0$),

$$\eta(x,t) = \frac{|\alpha_1|}{\alpha}A(s,\tau), \quad x = \sqrt{\frac{\alpha^2}{6\beta|\alpha_1|}}s, \quad t = \beta \cdot \left(\frac{\alpha^2}{6\beta|\alpha_1|}\right)^{3/2}\tau, \quad (33)$$

the Gardner equation (20) transforms to the canonical form

$$\frac{\partial\eta}{\partial t} + 6\eta(1 + p\eta)\frac{\partial\eta}{\partial x} + \frac{\partial^3\eta}{\partial x^3} = 0. \quad (34)$$

Parameter $p = \pm 1$ denotes the sign of the coefficient α_1. When it is positive ($p = 1$), we refer to equation (34) as GE+ and as GE− in the opposite case ($p = -1$). When $\alpha_1 = 0$, the Gardner soliton (20) transforms to the KdV soliton and is given by the formula:

$$A_s(s,\tau) = \frac{a}{\cosh^2\left(\sqrt{\dfrac{\alpha a}{12\beta}}(s - s_0 - V\tau)\right)}. \quad (35)$$

Here, the soliton velocity, $V = c + \alpha a/3$, is expressed through the soliton amplitude a. The soliton shapes are given in Fig. 3 for various signs of cubic nonlinearity in the normalized equation (34). The KdV soliton is shown in Fig. 3a by dotted lines. The KdV solitons have only one polarity (defined by the sign of the quadratic nonlinearity coefficient α). It is explicitly seen that a soliton becomes narrower when it is higher, thus intensive solitons have a broader spectrum and require the correct consideration of not only higher nonlinear effects but also higher-order dispersive effects. In the simple case of a two-layer fluid, the solitons are up-crested if the pycnocline is placed below the middle level and vice versa. The solitons of the GE retain the polarity of KdV solitons, but they are broader (see Fig. 3a). The amplitude of such solitons is bounded by the limiting value, $a_{\lim} = -\alpha_1/\alpha$ (in dimensionless variables $a_{\lim} = 1$). The limiting soliton has a flat crest and its slopes are shock-like waves often called *kinks*. In fact, kinks can be considered

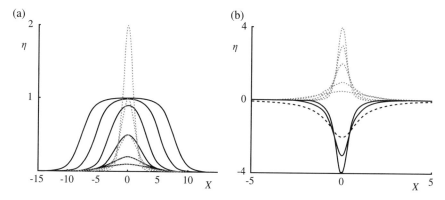

Figure 3: Solitons of different approximated models: (a) KdV solitons with amplitudes 0.1, 0.2, 0.5, 1, 2 (dotted lines) and GE− solitons with amplitudes 0.1, 0.2, 0.5, 0.9, 0.99, 0.999 (solid lines); (b) GE+ solitons with amplitudes 0.5, 1, 2, 3, 4, −2 (the algebraic soliton shown as the dashed line), −3, −4 The dimensionless form of equation (34) is used.

as independent nonlinear structures if the distance between them is large. This 'table' soliton plays an important role in wave dynamics, becoming a pedestal for other solitons, and changes significantly the process of disintegration of a pulse disturbance (as will be discussed in the next section).

The case $p = 1$ ($\alpha_1 > 0$) may be fulfilled, for instance, in the case of a three-layer stratification (Talipova et $al.$ 1999). The solitons remain similar to the KdV case, if they are positive (we suppose here α to be positive), see Fig. 3b. The soliton amplitude is not bounded in this model and a new solitary branch of opposite polarity exists. The latter have amplitudes larger (in modulus) than $a_{\text{alg}} = -2\alpha_1/\alpha$ ($a_{\text{alg}} = -2$ in the dimensionless variables). The soliton with amplitude a_{alg} has power-law tails (it is shown by the dotted line in Fig. 3b), and is called the algebraic soliton. The algebraic soliton is structurally unstable (Pelinovsky and Grimshaw 1997), but the other members of this branch demonstrate the usual vitality of solitons. When the solitons are very large (of either polarity), their dynamics is very close to the case of the modified KdV equation (when $\alpha \equiv 0$ in the Gardner equation).

Both the Gardner and the KdV equations are integrable, and the Cauchy problem can be solved exactly by analytical methods (see Lamb 1980 and Chapter 2). The solitons show an exceptional property of elastic collision among themselves and other waves. The multisoliton solutions can be obtained in an explicit form; in the case of the KdV equation, there are two kinds of soliton collisions. If the interacting solitons are very different (amplitude ratio $a_2/a_1 > 3$), then the fast one passes over the other and it is called overtaking collision (see Fig. 4a). If the solitons have close amplitudes, the process of exchange takes place: the fast soliton gets slower and gives its energy to the other one; there is always some distance between the solitons in such a type of collision (Fig. 4b), and they do not form a single-crested wave in contrast to the first type of interaction. It is obviously seen from Fig. 4 that the

(a) (b)

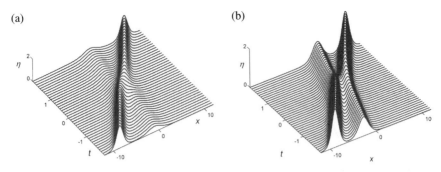

Figure 4: The processes of interaction of KdV solitons: (a) overtaking (the amplitude ratio is 4) and (b) exchange (the amplitude ratio is 2). The dimensionless form of equation (34) with $p = 0$ is used.

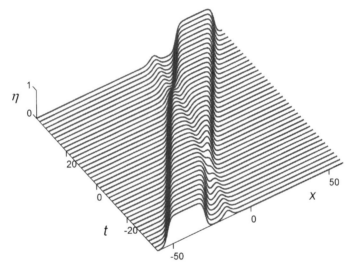

Figure 5: Collision of a soliton of amplitude 0.1 with a flat-crested soliton of
amplitude 0.999 within the framework of the GE− (34).

solitons get a phase shift during the collision, but their shapes and velocities remain
unchanged after the interaction.

When the cubic nonlinearity coefficient is negative ($p = -1$, the GE− equation),
the soliton amplitude varies from 0 to 1. The interaction of solitons with small and
moderate amplitudes is similar to the KdV case (Fig. 4). But if one soliton has
an amplitude close to the limiting value, the smaller soliton actually changes its
polarity climbing up onto the 'table' crest (Fig. 5). Each collision with the kinks
leads to a change of the polarity of the smaller soliton and a phase shift, but the
speed and the shape remain the same (Slyunyaev and Pelinovskii 1999).

In the case of the GE+ equation (the positive cubic nonlinearity), the interaction of the solitons of the same polarity is qualitatively the same as in the KdV

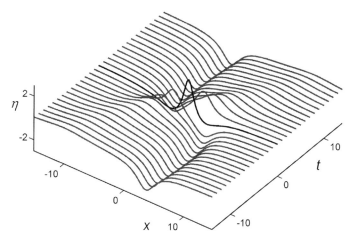

Figure 6: Collision of solitons of different polarities within the framework of the GE+ (34). The amplitudes are 0.2 (of the faster one; it goes from the left and is almost invisible when it does not interact with the second soliton) and -2.1 (of the slower one).

case (Slyunyaev 2001). The interaction of the solitons of opposite polarities is more complicated. Figure 6 gives such an example of the collision of two solitons with different polarities. The negative soliton is much larger than the positive one, which is almost invisible in the figure at the initial moment, but the larger wave almost changes its sign at the moment of the interaction (the solid line). A large phase shift is obtained by the negative soliton and the interaction is clearly seen in Fig. 6.

The discrete spectrum of the associated scattering problem defines localized solitary solutions of integrable equations (such as the KdV, see Chapter 2) or the Gardner equation (Miles 1979; Grimshaw et al. 2002a). The discrete spectrum of the KdV and GE− equations lies on an axis (the negative real axis for the KdV equation in the notation given in Chapter 2) of the spectral plane. The case of the GE+ equation turns out to be richer: the discrete spectrum may be complex (complex conjugated to describe real solutions). Pairs of complex conjugated spectral values define other localized nonlinear solutions of the GE+ equation – nonlinear wave packets or 'breathers' (Pelinovsky and Grimshaw 1997):

$$\eta = 2\frac{\partial}{\partial x}\tan^{-1}\left[\frac{l\cosh(\Psi)\cos(\theta) - k\cos(\Phi)\sinh(\varphi)}{l\sinh(\Psi)\sin(\theta) + k\sin(\Phi)\cosh(\varphi)}\right], \qquad (36)$$

where θ and φ are 'travelling' phases,

$$\theta = k(x - wt) + \theta_0, \quad \varphi = l(x - vt) + \varphi_0, \quad w = -k^2 + 3l^2, \quad v = -3k^2 + l^2.$$

In contrast to solitons, this wave is described by two important 'energetic' parameters

$$\Phi + i\Psi = \tan^{-1}[l + ik],$$

(a)   (b)

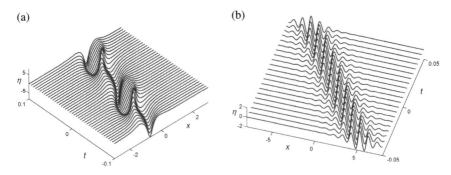

Figure 7: Breathers of the GE+.

and two initial phases (θ_0 and φ_0). A typical image of a breather is given in Fig. 7. A breather may look like the collision of two solitons of opposite polarities (Fig. 7a; compare with Fig. 6), or resemble a localized wave group containing a large number of individual waves (Figs 7b and 9b), which preserves its individuality. Although solitons always propagate faster than linear waves, a breather may move slower or faster depending on its parameters.

The Gardner equation remains a very simple model and is convenient for investigation since it is integrable. The Gardner, and even the KdV equation, may give a good approximation of many observed solitons in nature and in the laboratory. (see, for example, Ostrovsky and Stepanyants 1989, 2005). But these equations remain approximate and need assumptions of weak nonlinearity and dispersion to be satisfied. The fully nonlinear weakly dispersive models for interfacial waves in a two-layer fluid were discussed in Section 2. In fact, such models reduce to ordinary differential equations of the second order describing solitary waves, but cannot usually be solved analytically. Qualitatively, the soliton shape in the fully nonlinear models is the same as in the framework of the GE− (although it is valid for weakly nonlinear interfacial waves in a two-layer fluid): the solitary wave amplitude is bounded, and its exact value within the fully nonlinear model is (Choi and Camassa 1999)

$$a_{\lim} = \frac{h_2 - h_1\sqrt{\rho_2/\rho_1}}{1 + \sqrt{\rho_2/\rho_1}}, \tag{37}$$

and the speed of the limiting 'table' soliton is

$$c_{\lim}^2 = gh\frac{1 - \sqrt{\rho_2/\rho_1}}{1 + \sqrt{\rho_2/\rho_1}}, \quad h = h_1 + h_2. \tag{38}$$

In the Boussinesq limit (layers with near densities), the limiting value for the soliton amplitude is

$$a_{\lim} = (h_2 - h_1)/2, \tag{39}$$

and the speed of the 'table' soliton is

$$c_{\lim} = \frac{c_h}{2}, \quad c_h = \sqrt{g'h}. \tag{40}$$

So, the crest of the limiting soliton is situated exactly in the middle of the total water depth. The limiting amplitudes and velocities of solitons given by some models are compared in Fig. 8. Solid lines represent the case of the Gardner equation, dashed lines correspond to the extended Gardner equation (with the first four nonlinear terms taken from the fully nonlinear equation (24)), and crests represent the 'β-model' of Ostrovsky and Grue (2003). It is seen from Fig. 8a that the crests of limiting solitons are located near the middle of the fluid for a wide range of values of h_1 for all the models. If h_1 and h_2 differ much, the approximate models underestimate the limiting amplitude. Curves of limiting speeds of solitary waves look quite different for the Gardner and extended Gardner equations when h_1 and h_2 are very different (Fig. 8b). It is evident that the limiting velocity approaches the exact value $c_h/2$ when the model is improved. The dashed line in Fig. 8b shows the linear velocity c.

More sophisticated models may be used for the description of internal solitons (see Section 2), which improve the quantitative description of the dynamics and introduce new effects (the most important seems to be the inelasticity of the soliton interaction, generally due to the loss of integrability, but they are energetically small). An interesting approximate approach to the study of large-amplitude soliton interaction analysis has been developed by Gorshkov et al. (2004) based on soliton representation as a particle.

Actually when the pycnocline is relatively close to the middle of the fluid, the Gardner equation provides a good approximation of the fully nonlinear

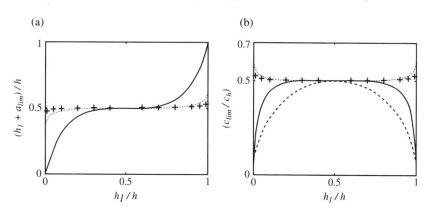

Figure 8: Position of the crest of the limiting soliton (a) and the limiting velocity (b) versus the normalized lower layer thickness. The solid line represents the Gardner equation, the dotted line represents the extended Gardner equation (when the first four nonlinear terms are taken into account in (24)) and crests show the β-model of Ostrovsky and Grue. The dashed line in (b) gives the linear velocity c.

models. The checking of observed and experimental internal solitons with different improved models, including fully nonlinear numerical simulations, may be found, for instance, in Mirie (1985), Michallet and Barthelemy (1998), Grue *et al.* (1999), Vlasenko *et al.* (2000), Ostrovsky and Grue (2003) and Small and Hornby (2005).

4 Evolution of initial disturbances

Usually, short-scale large-amplitude waves are generated by long-scale waves such as tides. This process may be considered theoretically as the Cauchy problem for the initial disturbance; it is classical if studied within the frameworks of the KdV equation (see, for instance, Lamb 1980). A negative initial disturbance of the Gardner equation (34) in the KdV limit ($p = 0$) evolves into a dispersive train, as shown in Fig. 9a, which may be represented via Painleve functions that are nonlinear analogues of the Airy function. Any positive finite disturbance evolves into one or many solitons and quasi-linear waves forming a decaying dispersive tail (Fig. 10a). Depending on the actual shape of the initial perturbation, the number of solitons and the distribution of energy between the solitary and dispersive parts of the waves may be different. Due to the property of integrability, the number of solitons is conserved in the process of wave propagation, although the dispersive tail vanishes with time. Thus, solitons represent the asymptotic behaviour of the wave field.

This bewitching process of soliton formation, and then the recurrence of initial state in the periodic domain, was observed first by Zabusky and Kruskal in 1965 and led to the discovery of the exceptional role of solitons in integrable systems. Due to the linear relation between amplitudes and velocities of KdV solitons, they are generated in order, see Fig. 10a. The largest (fastest) goes first and the slower and lower solitons follow it. The amplitude of the leading soliton may be almost twice the amplitude of the initial impulse.

The initial-value problem for the Gardner equation can also be solved exactly due to its integrability (Miles 1979; Grimshaw *et al.* 2002a). If the spectrum of the associated linear scattering problem is found, then it survives in time and the

(a)   (b)

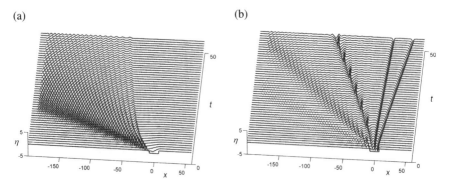

Figure 9: Evolution of the initial negative box-like impulse of amplitude 1.2 within the frameworks of (a) the KdV equation and (b) the GE+ equation.

(a)

(b)

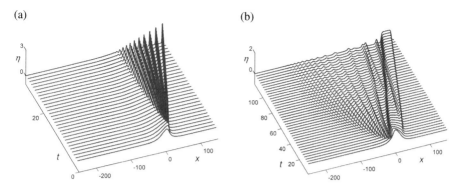

Figure 10: Evolution of the initial positive pulse of amplitude 1.2 within the frameworks of (a) the KdV equation and (b) the GE– equation. Note the
different scales of time and field in the figures.

result of the evolution may be formally found at any moment of time. The discrete
spectrum of the scattering problem corresponds to the localized solutions, such as
solitons and breathers (Lamb 1980).

Naturally, the low-amplitude limit of the GE shows qualitatively similar results
to the KdV evolution of an initial disturbance. This remains true for purely positive
disturbances of the GE+ equation and negative ones in the GE– equation. In
contrast to the KdV case, an intense negative impulse may evolve into a set of
solitons and breathers as shown in Fig. 9b, when the positive cubic nonlinearity is
taken into account (the GE+ case). Two negative solitons propagating to the right
are observed in the figure. Furthermore, breathers are also generated and two of
these may be seen clearly. One of the breathers contains few individual waves and
slowly propagates to the left in Fig. 9b, while the other contains a large number of
individual waves and moves to the left faster.

The transformation of initially positive intensive impulse disturbances happens
differently in the case of the KdV and the GE– equations (Fig. 10). In the latter case,
the appearance of the leading 'table' soliton is clearly seen, while the generation of
smaller-amplitude solitons is suppressed (Grimshaw *et al.* 2002a). Each slope of the
initial impulse generates solitary waves: one set of these has positive polarity and
forms behind the back slope of the initial impulse, but other solitons are generated
on the crest of the forming wide soliton behind the front of the impulse. Initially,
these solitons are negative and change their signs as they go down from the wide
soliton (this happens due to the larger velocity of the 'table' soliton, see Fig. 10b).
The process of collision of the wide soliton with a smaller one was described in
Section 3 and is shown in Fig. 5 and also in Fig. 10b. The presence of two groups
of solitons destroys the usual order of solitary waves, which collide according to
the difference in their speeds. Due to the limitation in soliton amplitude and speed
provided by the GE–, the generated solitons are typically less and slower (note the
different scales in Fig. 10a and b).

The knowledge of these scenarios gives a good qualitative idea about the processes of soliton generation. Disintegration of an internal bore-like disturbance is

often considered to lead to further evolution of internal waves; in Boussinesq-like and fully nonlinear models. This was studied by Lamb and Yan (1996), Vlasenko *et al.* (2000, 2002) and Lamb (2002). For example, the study of a certain oceanic area (Malin Shelf in the Atlantic) demonstrates a weak difference from the description given by the Gardner equation (Small and Hornby 2005).

5 Solitons in variable media

The multiscale asymptotic approach to the derivation of the KdV-type evolution equations takes into account the vertical structure of the internal wave fields through a modal representation. This simplification may turn out to be a strong point of the models of this type. Besides the first evident benefit (which is the simplicity of the models from an analytical and computational point of view – they are 1+1 dimensional), these models may also take into account the horizontal variability of the path of the internal waves (such as variable stratification, depth, etc). Such models may be derived after a suitable modification of the asymptotic technique considered in Section 2 (see Zhou and Grimshaw 1989; Pelinovsky *et al.* 1994; Holloway *et al.* 2002). They require the variation of environmental parameters along the wave path to be slow compared with the internal wave scale and result in variable coefficients of the evolution equations.

   The background conditions, indeed, change significantly during the internal solitary waves propagation over the continental shelf, which results in the transformation of solitary waves (Liu *et al.* 1998; Holloway *et al.* 2002; Zhao *et al.* 2003; Grimshaw *et al.* 2004). The most startling and most easily observed effect is the change of soliton polarity. This process is very often observed in the coastal zone, both directly and from satellite data (see, for instance, Zhao *et al.* 2003) and may be simply understood in terms of the KdV theory. Indeed, the soliton polarity is defined by the sign of the quadratic nonlinearity, which strongly depends on stratification and may change its sign (this is evident for the two-layer stratification, see Section 2). This term is the most important in the process of soliton transformation. The degeneration of the quadratic nonlinear term requires the consideration of a higher-order nonlinearity, leading to a generalized Gardner equation (Grimshaw *et al.* 1998, 1999).

$$\frac{\partial A}{\partial T} + c\frac{\partial A}{\partial X} + \alpha(T)A\frac{\partial A}{\partial T} + \alpha_1 A^2\frac{\partial A}{\partial T} + \beta\frac{\partial^3 A}{\partial T^3} = 0 \qquad (41)$$

represents a first-order model describing this process, where all the coefficients except α can be regarded as being constant and $\alpha = 0$ at the 'turning' point. This equation is not integrable, but conserves the mass

$$M = \int_{-\infty}^{\infty} A(X, T)\mathrm{d}X = \text{constant} \qquad (42)$$

and energy (momentum) integrals

$$E = \frac{1}{2}\int_{-\infty}^{\infty} A^2(X, T)\mathrm{d}X = \text{constant.} \qquad (43)$$

These conserved quantities may be used for the description of the adiabatic stage of the soliton evolution (Grimshaw *et al.* 1999; Nakoulima *et al.* 2004). When the soliton polarity is defined by the sign of the quadratic nonlinearity (the GE− case or the low-amplitude limit of GE+), a soliton cannot pass the turning point, adiabatically changing its parameters and preserving its own identity, instead it undergoes drastic changes. Then a small-amplitude soliton transforms into a dispersive wave packet in the vicinity of the turning point, and a negative soliton of a lesser amplitude is formed later (see Fig. 11a). If a wide ('table') soliton passes the turning point, the generated soliton of opposite polarity is again wide, but its mass integral is less due to the radiation of the dispersive tail (Fig. 11b).

In the case $\alpha_1 > 0$ (the GE+ equation), cubic nonlinearity supports intensive solitons of both polarities (this is close to the case of modified KdV equation with the only nonlinear term being the cubic nonlinear term, that is $\alpha \equiv 0$) resulting in a relatively small change of the soliton shape and amplitude while passing the transition zone, as shown in Fig. 12. If the initial soliton is small, its transformation is similar to that shown in Fig. 11.

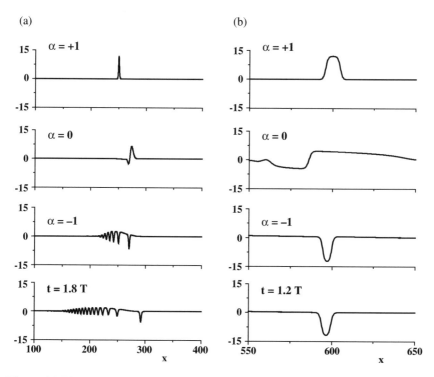

Figure 11: Transformation of a soliton passing the turning point: small-amplitude case (a) and a wide soliton (b) (numerical simulation). The model is the Gardner equation (41) with $c = 0$, $\alpha = 1 - t/T$, $\alpha_1 = -0.08$ and $\beta = 1$.

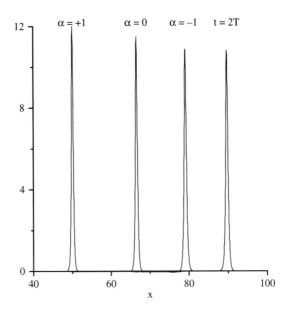

Figure 12: Transformation of an intensive soliton passing the turning point. The model is the Gardner equation (41) with $c = 0$, $\alpha = 1 - t/T$, $\alpha_1 = 1$ and $\beta = 1$.

Breathers may also be generated by a soliton due to change of the nonlinear coefficients of the Gardner equation, and if $\alpha_1 > 0$ after the turning point (Grimshaw *et al.* 1999; Nakoulima *et al.* 2004), the effect of soliton transformation becomes multifarious.

The opposite case of a sudden change of environmental parameters may also be considered within these models. Then the Cauchy problem is solved (see Section 4) where the wave field just before the transition zone gives the initial conditions, and the coefficients of the equation are changed to the new (after the transition) values. Typically, this leads to the so-called soliton fission.

6 Conclusion

Internal solitons are fascinating, nonlinear geophysical objects that allow detailed investigation using sophisticated measuring devices developed in recent years. Their most extensive theoretical description is based on the solution of the primitive, fully three-dimensional equations of hydrodynamics with the essential consideration of vertical stratification due to the salinity, temperature distribution and inhomogeneous shear flows. It is a very expensive approach that needs a powerful computer and a lot of information about the environmental conditions, and may be difficult to interpret. At present, only simple model simulations are being carried out with this approach. Another common approach to study these internal waves considers the modal structure supported by the stratification (the modes may be obtained through numerical solution of the appropriate boundary-value problem, or simplified stratifications, such as the two-layer fluid, may be employed leading to analytical solutions) and evolution equations that describe the wave motion in the horizontal plane. This approach is less expensive from the point of view of numerical computations and uses averaged information about the stratification (thus, is less affected by the sea variability and errors of measurements). The nonlinear evolution equations describing internal waves may be simple (and even integrable), but the quality of the description of the internal waves is reasonable and, often, no less accurate than natural measurements. The KdV equation is the first, and a popular, model of nonlinear internal waves since 1966 (Benney 1966). Considering the real variability of sea stratification, this equation should be replaced by the Gardner equation, which is supposed to be the simplest model describing internal solitons. The evolution equation may be improved by introducing other terms of higher orders, using modified asymptotic techniques or employing specific stratifications that allow one to formulate a highly (fully) nonlinear theory; heuristic methods may also be useful in this way. These more accurate models usually show qualitatively similar results and the difference may be confidently observed only in laboratory experiments.

On the basis of the Gardner equation and its generalizations, the description of the observed internal solitons on the ocean shelves becomes possible. This model includes the smooth horizontal variability of the stratification with ease (which is a strong effect on a sloping shelf). Ray methods allow one to take into account the second horizontal dimension (the wave refraction), while, along a ray, the variable-coefficient Gardner equation is solved. The coefficients may be found on the basis of the available atlas of sea stratification. The existence of exact solutions of the integrable version of the equation helps to verify the numerical model and interpret the obtained results. All these above-mentioned facts allow us to get a realistic description of the measured internal solitons over the shelves within this relatively simple model.

Acknowledgements

This work was supported by grants INTAS 03-51-3728, 06-1000013-9236, RFBR 04-05-39000, 05-05-64333, 06-05-64232 and programmes for young researchers MK-1358.2005.1 (for O.P.) and the Russian Science Support Foundation (for A.S.).

References

Akylas, T. & Grimshaw, R., Solitary internal waves with oscillatory tails. *J. Fluid Mech.*, **242**, pp. 279–298, 1992.

Baines, P.G., *Topographic Effects in Stratified Flows*, Cambridge University Press: Cambridge, 1995.

Benney, D.J., Long nonlinear waves in fluid flows. *J. Math. Phys.*, **45**, pp. 52–63, 1966.

Cheung, T.K. & Little, C.C., Meteorological tower, microbarograph array, and radar observations of solitary-like waves in the nocturnal boundary layer. *J. Atmos. Sci.*, **47**, pp. 2516–2536, 1990.

Choi, W. & Camassa, R., Fully nonlinear internal waves in a two-fluid system. *J. Fluid Mech.*, **396**, pp. 1–36, 1999.

Craig, W., Guyenne, P. & Kalisch, H., A new model for large amplitude long internal waves. *C. R. Mecanique*, **332**, pp. 525–530, 2004.

Dubriel-Jacotin, L., Sur les ondes type permanent dans les liquids heterogenes. *Atti della reale Academic Nationalite dci Lincei*, **15**, pp. 44–52, 1932.

Engelbrecht, J.K., Fridman, V.E. & Pelinovsky, E.N., *Nonlinear Evolution Equations (Pitman Research Notes in Mathematics Series, No. 180)*, Longman: London, 1988.

Fokas, A., Grimshaw, R.H.J. & Pelinovsky, D.E., On the asymptotic integrability of a higher-order evolution equation describing internal waves in a deep fluid. *J. Math. Phys.* **37**, pp. 3415–3421, 1996.

Global Ocean Associates, An atlas of oceanic internal solitary-like waves and their properties, 2004, http://www.internalwaveatlas.com

Gorshkov, K.A., Ostrovsky, L.A., Soustova, I.A. & Irisov, V.G., Perturbation theory for kinks and its application for multisoliton interactions in hydrodynamics. *Phys. Rev. E.*, **69**, 016614, 2004.

Green, A.E. & Naghdi, P.M., A derivation of equations for wave propagation in water of variable depth. *J. Fluid Mech.*, **78**, pp. 237–246, 1976.

Grimshaw, R., Internal solitary waves. *Environmental Stratified Flows*, ed. R. Grimshaw, Kluwer: Boston, pp. 1–29, 2002.

Grimshaw, R., Pelinovsky, E. & Talipova, T., Solitary wave transformation due to a change in polarity. *Stud. Appl. Math.*, **101**, pp. 357–388, 1998.

Grimshaw, R., Pelinovsky, E. & Talipova, T., Solitary wave transformation in a medium with sign-variable quadratic nonlinearity and cubic nonlinearity. *Physica D*, **132**, pp. 40–62, 1999.

Grimshaw, R., Pelinovsky, D., Pelinovsky, E. & Slunyaev, A., Generation of large-amplitude solitons in the extended Korteweg–de Vries equation. *Chaos*, **12**, pp. 1070–1076, 2002a.

Grimshaw, R., Pelinovsky, E. & Poloukhina, O., Higher order Korteweg – de Vries models for internal solitary waves in a stratified shear flow with a free surface. *Nonlin. Proc. Geoph.,* **9**, pp. 221–235, 2002b.

Grimshaw, R., Pelinovsky, E., Talipova, T. & Kurkin, A., Simulation of the transformation of internal solitary waves on oceanic shelves. *J. Phys. Oceanography*, **34**, pp. 2774–2791, 2004.

Grue, J., Jensen, A., Rusas, P.O. & Sveen, J.K., Properties of large-amplitude internal waves. *J. Fluid Mech.*, **380**, pp. 257–278, 1999.

Helfrich, K.R. & Melville, W.K., Long nonlinear internal waves. *Annu. Rev. Fluid Mech.*, **38**, pp. 395–425, 2006.

Holloway, P., Pelinovsky, E. & Talipova, T., Internal tide transformation and oceanic internal solitary waves. *Environmental Stratified Flows*, ed. R. Grimshaw, Kluwer: Boston, pp. 29–60, 2002.

Jeans, D.R.G., *Solitary Internal Waves in the Ocean: A Literature Review Completed as Part of the Internal Wave Contribution to Morena*, UCES, Marine Science Labs, University of North Wales, Rep. U-95, 1995.

Kakutani, T. & Yamasaki, N., Solitary waves in a two-layer fluid. *J. Phys. Soc. Japan*, **45**, pp. 674–679, 1978.

Kodama, Y., Nearly integrable systems. *Physica D*, **16**, pp. 14–26, 1985.

Lamb, G.L., Jr., *Elements of Soliton Theory*, Wiley: New York, 1980.

Lamb, K., A numerical investigation of solitary internal waves with trapped cores formed via shoaling. *J. Fluid Mech.*, **451**, pp. 109–144, 2002.

Lamb, K. & Yan, L., The evolution of internal wave undular bores: comparisons of a fully nonlinear numerical model with weakly nonlinear theory. *J. Phys. Ocean.*, **26**, pp. 2712–2734, 1996.

Lee, C. & Beardsley, R.C., The generation of long nonlinear internal waves in a weakly stratified shear flow. *J. Geophys. Research*, **79**, pp. 453–462, 1974.

Liu, A.K., Chang, Y.S., Hsu, M.K. & Liang, N.K., Evolution of nonlinear internal waves in the East and South China Seas. *J. Geoph. Res.*, **103**, pp. 7995–8008, 1998.

Madsen, P.A., Bingham, H.B. & Schäffer, H.A., Boussinesq-type formulations for fully nonlinear and extremely dispersive water waves: derivation and analysis. *Proc. R. Soc. Lond. A*, **459**, pp. 1075–1104, 2003.

Michallet, H. & Barthelemy, E., Experimental study of interfacial solitary waves. *J. Fluid Mech.*, **366**, pp. 159–177, 1998.

Miles, J.W., On the stability of heterogeneous shear flows. *J. Fluid Mech.*, **10**, pp. 496–509, 1961.

Miles, J.W., On internal solitary waves. *Tellus*, **31**, pp. 456–462, 1979.

Miles, J.W., On internal solitary waves II. *Tellus*, **33**, pp. 397–401, 1981.

Mirie, R.M., An internal solitary wave with bounded amplitude. *J. Phys. Soc. Japan*, **54**, pp. 3332–3336, 1985.

Miyata, M., Long internal waves of large amplitude. *Nonlinear Water Waves*, ed. K. Horikawa and H. Maruo, Springer–Verlag: Berlin, pp. 399–406, 1988.

Nakoulima, O., Zahibo, N., Pelinovsky, E., Talipova, T., Slunyaev, A. & Kurkin, A., Analytical and numerical studies of the variable-coefficient Gardner equation. *Appl. Math. Comp.*, **152**, pp. 449–471, 2004.

Ostrovsky, L.A., Nonlinear internal waves in a rotating ocean. *Oceanology*, **18**, pp. 119–125, 1978.

Ostrovsky, L.A. & Grue, J., Evolution equations for strongly nonlinear internal waves. *Physics of Fluids*, **15**, pp. 2934–2948, 2003.

Ostrovsky, L.A. & Stepanyants, Yu.A., Do internal solitons exist in the ocean? *Rev. of Geophysics*, **27(3)**, pp. 293–310, 1989.

Ostrovsky, L. & Stepanyants, Yu., Internal solitons in laboratory experiments: comparison with theoretical models. *Chaos*, **15**, 037111, 2005.

Pelinovsky, D. & Grimshaw, R., Structural transformation of eigenvalues for a perturbed algebraic soliton potential. *Phys. Lett. A.*, **229**, pp. 165–172, 1997.

Pelinovsky, E.N., Stepanyants, Yu.A. & Talipova, T.G., Modeling of internal wave propagation into the horizontally heterogeneous ocean. *Izvestiya, Atmospheric and Ocean Physics*, **30(1)**, pp. 79–85, 1994.

Peregrine, D.H., Long waves on a beach. *J. Fluid Mech.*, **27**, pp. 815–827, 1967.

Rottman, J.W. & Grimshaw, R., Atmospheric internal solitary waves. *Environmental Stratified Flows*, ed. R. Grimshaw, Kluwer: Boston, pp. 63–90, 2002.

Sabinin, K. & Serebryany, A., Intense short-period internal waves in the ocean. *J. Marine Res.*, **63**, pp. 227–261, 2005.

Slyunyaev, A.V., Dynamics of localized waves with large amplitude in a weakly dispersive medium with a quadratic and positive cubic nonlinearity. *JETP*, **92**, pp. 529–534, 2001.

Slyunyaev, A. & Pelinovskii, E., Dynamics of large-amplitude solitons. *JETP*, **89(1)**, pp. 173–181, 1999.

Slunyaev, A., Pelinovsky, E., Poloukhina, O. & Gavrilyuk, S., The Gardner equation as the model for long internal waves. *Proc. of Int. Symp. on Topical Problems of Nonlinear Wave Physics*, pp. 368–369, 2003.

Small, R.J. & Hornby, R.P., A comparison of weakly and fully non-linear models of the shoaling of solitary internal waves. *Ocean Modelling*, **8**, pp. 395–416, 2005.

Small, J., Sawyer, T.C. & Scott, J.C., The evolution of an internal bore at the Malin shelf break. *Ann. Geophysicae*, **17**, pp. 547–565, 1999.

Talipova, T., Pelinovsky, E., Lamb, K., Grimshaw, R. & Holloway, P., Cubic effects at the intense internal wave propagation. *Doklady Earth Sciences*, **365**, pp. 241–244, 1999.

Vlasenko, V., Brandt, P. & Rubino, A., Structure of large-amplitude internal solitary waves. *J. Phys. Oceanography*, **30**, pp. 2172–2185, 2000.

Vlasenko, V.I. & Hutter, K., Transformation and disintegration of strongly nonlinear internal waves by topography in stratified lakes. *Annales Geophysicae*, **20**, pp. 2087–2103, 2002.

Vlasenko, V., Stashchuk, N. & Hutter, K., *Baroclinic Tides: Theoretical Modeling and Observational Evidence*, Cambridge University Press: Cambridge, 2005.

Whitham, G.B., *Linear and Nonlinear Waves*, Wiley: New York, 1974.

Zabusky, N.J. & Kruskal, M.D., Interaction of solutions in a collisionless plasma and recurrence of initial states. *Phys. Rev. Lett.*, **15**, pp. 240–243, 1965.

Zhao, Zh., Klemas, V., Zheng, Q. & Yan, X.-H., Satellite observation of internal solitary waves converting polarity. *Geoph. Res. Lett. 2003*, **30(19)**, 1988.

Zhou, X. & Grimshaw, R., The effect of variable currents on internal solitary waves. *Dyn. Atm. Oceans*, **14**, pp. 17–39, 1989.

CHAPTER 5

Solitary waves in rotating fluids

T.R. Akylas

Department of Mechanical Engineering, Massachusetts Institute of Technology, Cambridge, MA, USA.

Abstract

In this chapter, we discuss solitary waves when a swirl component is present in the background flow. Depending on the particular physical setting, rotation is responsible for inertial solitary waves but may significantly affect the propagation of gravity solitary waves as well. Here, two specific examples that illustrate this dual role of rotation are considered: (1) axisymmetric inertial waves in a rotating homogeneous fluid in a rigid tube; (2) three-dimensional internal gravity waves in a stratified fluid layer in a rotating channel. In (1), the propagation of weakly nonlinear long-wave disturbances is typical of a waveguide; each inertial wave mode is governed by a Korteweg–de Vries (KdV) equation to leading order, and solitary waves with sech^2 profiles along the tube axis arise. In the special case of uniform rotation, it is also possible to construct finite-amplitude solitary waves analytically, and these solutions recently have been linked to the phenomenon of vortex breakdown. In (2), on the other hand, rotation primarily modifies the transverse structure of internal solitary waves propagating along the channel. When the rotation is relatively weak, in particular, so that the Rossby radius of deformation is comparable to the transverse wave scale, three-dimensional long-wave disturbances are governed by a rotation-modified Kadomtsev–Petviashvili (KP) equation. Consistent with laboratory observations, this equation predicts that an initially straight-crested KdV solitary wave experiences radiation damping as it propagates along the channel, and its wave crest is curved backwards owing to the shedding of Poincaré waves behind.

1 Introduction

Wave motion is intimately connected with the existence of a restoring mechanism, via which disturbances to an equilibrium or steady-motion state are brought back towards the unperturbed state. In fluid flows, familiar examples are gravity, which is responsible for free-surface and internal gravity waves; surface tension that becomes important for surface waves of less than a few centimeter wavelength; and compressibility, which gives rise to sound waves. As discussed in other chapters, these types of wave motion support solitary waves under some circumstances.

Here, we are concerned with the propagation of solitary waves in the presence of a swirl component in the background flow. While the restoring force responsible for wave motion in this instance may not be immediately obvious, the situation is analogous to that of internal gravity waves in a stably stratified incompressible fluid, since, according to Kelvin's theorem, circulation in inviscid rotating flow is conserved similarly to the density following a fluid particle in stratified flow. In the latter case, a particle displaced vertically from hydrostatic equilibrium will tend to return towards equilibrium if the background density decreases with height. Similarly, in rotating flow, a fluid ring that is displaced radially will be brought back towards its steady state if the background swirl is such that the square of the circulation increases with radius (see, for example, Yih 1979). As a result, inertial waves in a rotating fluid share many common characteristics with gravity internal waves, and there are various other instances of analogous behavior between rotating and stratified flows (Veronis 1970).

Apart from acting as the driving mechanism of inertial waves, rotation may also impact the propagation of surface and internal gravity waves. As these waves are quite common in oceans, lakes and the atmosphere, the earth's rotation could play a significant part (see Grimshaw *et al.* 1998b for a comprehensive review). In geophysical applications, typically the Coriolis parameter is much smaller than the buoyancy frequency; rotation nevertheless can have a cumulative effect on disturbances, like solitary waves, that may propagate stably for long distances and times.

In this chapter, the discussion focuses on two specific physical problems where rotation is central to the propagation of solitary waves. The first deals with axisymmetric inertial waves in a rotating homogeneous fluid in a rigid tube, a problem of some relevance to the understanding of the phenomenon of vortex breakdown (see, for example, the review by Leibovich 1991). This flow configuration is typical of a waveguide, and the analysis is analogous to that of internal gravity waves in a stratified fluid layer bounded by rigid walls. The second example examines the role of relatively weak rotation in the evolution of internal gravity solitary waves in a shallow channel of finite width, an issue of interest in a geophysical context as noted earlier.

2 Axisymmetric inertial solitary waves

A long, circular, rigid tube is filled with an inviscid incompressible fluid of constant density. It is convenient to use polar coordinates (x, r, θ) and, for axisymmetric flow, to work with a stream function $\psi(x, r, t)$ so that

$$u = \frac{1}{r}\psi_r, \quad v = -\frac{1}{r}\psi_x, \tag{1}$$

u and v denoting the axial and radial velocity components, respectively. The continuity equation is thus automatically satisfied and eliminating the pressure from the axial- and radial-momentum equations yields

$$\zeta_t + u\zeta_x + v\zeta_r - \frac{v\zeta}{r} - 2\frac{ww_x}{r} = 0, \tag{2}$$

$\zeta = v_x - u_r$ being the azimuthal vorticity and w the azimuthal velocity.

It also follows from the azimuthal-momentum equation that

$$\Gamma_t + u\Gamma_x + v\Gamma_r = 0, \tag{3}$$

confirming that the circulation $\Gamma = rw$ is conserved, in accordance with Kelvin's theorem.

Consider now a base flow consisting of a steady swirl $w_0(r)$ with circulation $\Gamma_0 = rw_0$, on which are superposed small-amplitude long-wave perturbations:

$$\psi \to \epsilon\psi(X, r, T), \quad \Gamma = \Gamma_0(r) + \epsilon\gamma(X, r, T). \tag{4}$$

Here, $0 < \epsilon \ll 1$ measures the perturbation amplitude and controls the nonlinear effects, while $(X, T) = \mu(x, t)$ are scaled variables associated with long waves in the axial direction, $0 < \mu \ll 1$ being the long-wave parameter (the ratio of the pipe radius to the axial lengthscale of the perturbation).

Substituting (4) in equations (2) and (3) and taking into account (1), it is found that

$$\zeta_T - 2\frac{\Gamma_0}{r^3}\gamma_X = 2\epsilon\frac{\gamma\gamma_X}{r^3} - \epsilon J(\zeta/r, \psi), \tag{5}$$

$$\gamma_T - \frac{\Gamma_0'}{r}\psi_X = -\frac{\epsilon}{r}J(\gamma, \psi), \tag{6}$$

where

$$\zeta = -\frac{1}{r}\left(\psi_{rr} - \frac{1}{r}\psi_r\right) - \mu^2\frac{\psi_{XX}}{r} \tag{7}$$

and $J(a, b) = a_X b_r - a_r b_X$ stands for the Jacobian. Equations (5) and (6), combined with (7), form a system to be solved for ψ and γ; in addition, ψ must satisfy the boundary conditions

$$\psi = 0 \quad (r = 0, 1), \tag{8}$$

requiring the radial velocity to vanish at the center and on the wall of the pipe.

Setting $\epsilon = \mu = 0$ in (5), (6) and (7), the leading order problem for ψ is

$$\left(\psi_{rr} - \frac{1}{r}\psi_r\right)_{TT} + \frac{(\Gamma_0^2)'}{r^3}\psi_{XX} = 0, \tag{9}$$

subject to the boundary conditions (8). Equation (9) admits separable solutions

$$\psi = A(X - cT)f(r),\tag{10}$$

where

$$f_{rr} - \frac{1}{r}f_r + \frac{(\Gamma_0^2)'}{c^2 r^3}f = 0 \quad (0 < r < 1)\tag{11}$$

with

$$f = 0 \quad (r = 0, 1),\tag{12}$$

in view of (8).

The boundary-value problem in (11) and (12) is a singular Sturm–Liouville problem, $\lambda^2 = 1/c^2$ being an eigenvalue parameter. Note that Γ_0^2, the square of the circulation associated with the basic swirl, appears explicitly in (11), and the sign of $(\Gamma_0^2)'$ decides whether wave propagation is possible, as suggested on physical grounds earlier. Specifically, if Γ_0^2 increases with r so $(\Gamma_0^2)' > 0$, all eigenvalues are positive, $0 < \lambda_1 < \lambda_2 < \ldots$ and $\lambda_n^2 \to \infty$ as $n \to \infty$, implying that the corresponding modes $\{f_n\}$ $(n = 1, 2, \ldots)$ are propagating $(c_n^2 > 0)$. In general, this eigenvalue problem has to be solved numerically, although in some special cases (e.g. when $w_0 = \Omega_0 r$, corresponding to uniform rotation) analytical solution is possible.

It is now clear that a swirling flow $w_0(r)$ with $(\Gamma_0^2)' > 0$ in a tube acts as a waveguide, similar to a stratified fluid layer between rigid walls. In the long-wave limit ($\mu \to 0$), there is an infinite set of linear modes, each propagating without dispersion along the tube with the corresponding long-wave speed. For problems of this type, Benney (1966) devised a systematic expansion procedure of deriving amplitude–evolution equation for the waveguide modes, accounting for nonlinear and dispersive effects at each order. The procedure establishes the Korteweg–de Vries (KdV) equation as the canonical weakly nonlinear long-wave evolution equation for the unidirectional propagation of each mode, to leading order in nonlinearity and dispersion, and solitary waves with sech^2 profile arise. Out of these KdV solitary wave solutions, however, only the one corresponding to the first mode ($n = 1$) is expected to be truly locally confined, as the rest are likely to feature lower-mode short-scale oscillatory tails, a phenomenon beyond all orders of the long-wave expansion (Akylas and Grimshaw 1992).

The expansion procedure of Benney (1966) was first applied to rotating flow by Leibovich (1970) and now we describe the salient features of the analysis. Returning to the boundary-value problem (5)–(8), we assume that $\epsilon = \mu^2$; this choice ensures that nonlinear and dispersive effects are equally important. We then introduce the expansions

$$\psi = \psi^{(0)} + \epsilon\psi^{(1)} + \epsilon^2\psi^{(2)} + \cdots, \quad \gamma = \gamma^{(0)} + \epsilon\gamma^{(1)} + \epsilon^2\gamma^{(2)} + \cdots,\tag{13}$$

the leading order solution being a single long-wave mode, $n = M$,

$$\psi^{(0)} = A(\xi, \tau) f_M(r), \quad \gamma^{(0)} = -\frac{A(\xi, \tau)}{c_M} \frac{\Gamma_0'}{r} f_M(r). \tag{14}$$

Here $\xi = X - c_M T$, corresponding to a reference frame moving with the long-wave speed c_M, and $\tau = \epsilon T$ is a scaled time, anticipating that the effects of weak nonlinearity and dispersion come into play after a long time, $T = O(1/\epsilon)$.

The amplitude $A(\xi, \tau)$ remains unspecified to leading order in expansions (13). Benney (1966) noted, however, that, to have the flexibility to eliminate secular terms at later stages in the expansion procedure, $A(\xi, \tau)$ must satisfy a KdV-type evolution equation of the form:

$$A_\tau = \beta A A_\xi + \delta A_{\xi\xi\xi} + O(\epsilon). \tag{15}$$

The coefficients β and δ in this amplitude equation are determined by imposing solvability conditions on the inhomogeneous problems for the $O(\epsilon)$ corrections $\psi^{(1)}$ and $\gamma^{(1)}$ in (13).

Using the leading order solution (14) to evaluate the $O(\epsilon)$ terms in (5) and (6) reveals that $\psi^{(1)}$ and $\gamma^{(1)}$ take separable forms as well:

$$\psi^{(1)} = A^2 f^{(1,0)}(r) + A_{\xi\xi} f^{(0,1)}(r), \tag{16a}$$

$$\gamma^{(1)} = A^2 q^{(1,0)}(r) + A_{\xi\xi} q^{(0,1)}(r), \tag{16b}$$

where $q^{(1,0)}$ and $q^{(0,1)}$ are known in terms of $f^{(1,0)}, f^{(0,1)}$ and f_M. Moreover, $f^{(1,0)}$ and $f^{(0,1)}$ satisfy inhomogeneous boundary-value problems of the form:

$$c_M^2 \left(f_{rr} - \frac{1}{r} f_r \right) + \frac{(\Gamma_0^2)'}{r^3} f = R \quad (0 < r < 1), \tag{17a}$$

$$f = 0 \quad (r = 0, 1), \tag{17b}$$

where R denotes a known forcing term, namely:

$$R^{(1,0)} = \frac{r^2}{2c_M} \frac{1}{(\Gamma_0^2)'} \left[((\Gamma_0^2)')^2 \right]' f_M^2 - \frac{\beta}{2c_M} \frac{(\Gamma_0^2)'}{r^3} f_M, \tag{18a}$$

$$R^{(0,1)} = - \left\{ \frac{(\Gamma_0^2)'}{c_M r^3} \delta + c_M^2 \right\} f_M. \tag{18b}$$

As the corresponding homogeneous problem has a nontrivial solution (the long-wave mode f_M), each of these problems has a solution if the forcing satisfies

$$\int_0^1 \frac{dr}{r} R f_M(r) = 0. \tag{19}$$

Imposing these solvability conditions to $R^{(1,0)}$ and $R^{(0,1)}$ in (18) specifies the coefficients β and δ in the KdV equation (15):

$$I\beta = \frac{1}{c_M^3} \int_0^1 \frac{r}{(\Gamma_0^2)'} \left[((\Gamma_0^2)')^2 \right]' f_M^3 \, dr, \tag{20}$$

$$I\delta = - \int_0^1 \frac{f_M^2}{r} \, dr, \tag{21}$$

with

$$I = \frac{2}{c_M^3} \int_0^1 \frac{(\Gamma_0^2)'}{r^4} f_M^2 \, dr. \tag{22}$$

It is worth noting that the coefficient β multiplying the nonlinear term of the KdV equation vanishes when the background flow has uniform angular velocity, $w_0 = \Omega_0 r$, as each linear long-wave mode happens to be also a nonlinear solution under these special flow conditions. An entirely analogous situation arises in a uniformly stratified fluid layer (constant buoyancy frequency) in the Boussinesq approximation, and Benney and Ko (1978) constructed finite-amplitude solitary wave solutions, similar to those of the KdV equation, by including small non-Boussinesq effects. The effects of transience and forcing on these large-amplitude internal solitary waves were later discussed by Grimshaw and Yi (1991), who pointed out that the transient term of the appropriate evolution equation is of the integral type and involves a nonlinear kernel.

In axisymmetric rotating flows, the properties of large-amplitude solitary waves, when the upstream flow is close to that of uniform axial flow and uniform rotation, were studied by Derzho and Grimshaw (2002). When the wave amplitude exceeds the critical value for incipient flow reversal, these solitary waves feature recirculation zones and may be relevant to the phenomenon of vortex breakdown.

Resonant axisymmetric flow of a rotating fluid past an obstacle on the axis of a cylindrical tube, in the general case that the upstream flow features a swirling and an axial component, was studied by Grimshaw (1990). As expected, close to resonance (when a certain long-wave mode happens to be nearly stationary in the reference frame of the obstacle), the response is governed by a forced KdV equation, and solitary waves of the resonant mode are radiated periodically upstream, as in the case of resonant free surface (Akylas 1984) and internal gravity waves (Grimshaw and Smyth 1986) over topography.

3 Internal gravity waves in a rotating channel

Maxworthy (1983) first drew attention to the effects of rotation on the dynamics of internal solitary waves. His laboratory experiments were conducted in a rotating channel with a stratified fluid layer and revealed that, while the solitary wave profile retains the familiar sech² shape along the channel, the wave amplitude varies exponentially in the transverse direction and, most surprisingly, the wave crest is curved backwards. The latter feature must be attributed to nonlinear effects, as linear Kelvin waves in a rotating channel have exponentially decaying amplitude but straight crests across the channel, and Maxworthy (1983) attempted to explain the curving of the solitary wave crest based on the dependence of the wave speed on amplitude. Later, Renouard *et al.* (1987) carried out similar experiments with two-layer flow in a long channel rotating on a large platform. Apart from confirming the earlier observations, they also noted that a solitary wave is accompanied by a train of small-amplitude waves trailing the main disturbance; moreover, the solitary wave amplitude is attenuated as the wave propagates along the channel, and this decay was attributed to viscous dissipation.

A systematic theoretical study of weakly nonlinear long waves in a rotating stratified fluid layer was made by Grimshaw (1985). It turns out that the effect of rotation is controlled by the parameter

$$v = \frac{\Omega l}{c_0}, \tag{23}$$

the ratio of l, the characteristic wavelength along the channel, to the Rossby radius of deformation, c_0/Ω (Ω is the angular velocity and c_0 the long-wave speed). When $v \gg \mu$ ($\mu = h/l$ denotes the long-wave parameter, h being the fluid depth), the wave evolution still is governed by a KdV equation, and the principal effect of rotation is an exponential decay of the wave amplitude in the transverse direction, as in the linear response. However, when Ω is slow enough so

$$v = \beta\mu, \quad \beta = O(1), \tag{24}$$

the effects of weak rotation are as important as those of nonlinearity, dispersion and transverse variation. In this distinguished limit, the KdV equation is replaced by a rotation-modified Kadomstev–Petriashvili (KP) equation that reflects a balance of all these weak effects. Note that the transverse structure of a solitary wave now is not known a priori, but has to be determined by solving this equation.

For a formal derivation of the rotation-modified KP equation, the reader is referred to Grimshaw(1985). Here, following Katsis and Akylas (1987), we sketch an intuitive derivation in the context of the two-layer configuration used in the experiments of Renouard *et al.* (1987).

The linear dispersion relation of waves with wavenumber k and frequency ω on the interface of two fluids (depths h_1 and h_2 and densities $\rho_1 < \rho_2$), bounded by rigid walls and rotating about the vertical with angular velocity Ω, is given by

$$\omega^2 = \frac{k}{\mu s} \frac{h_2}{h} \frac{\tanh q_1 \ \tanh q_2}{\tanh q_1 + (1 - \sigma)\tanh q_2}. \tag{25}$$

Here, we have used dimensionless variables, based on the typical wavelength l and the long-wave speed $c_0 = (g\sigma\bar{h})^{1/2}$, and the notation

$$q_1 = \frac{h_1}{h_2} \frac{\mu k}{s}, \quad q_2 = \frac{\mu k}{s}, \quad s = \left(1 - 4\frac{v^2}{\omega^2}\right)^{1/2}, \tag{26}$$

with

$$\mu = \frac{h_2}{l}, \quad \bar{h} = \frac{h_1 h_2}{h_1 + h_2}, \quad \sigma = \frac{\rho_2 - \rho_1}{\rho_2}, \tag{27}$$

g being the gravitational acceleration and ν the rotation parameter already defined in (23).

For long waves and slow rotation, the parameters μ and ν are small, and since the two fluid layers in the experiments of Renouard *et al.* (1987) had almost equal densities, $\sigma \ll 1$. Hence, retaining the leading-order dispersive and rotation effects, the dispersion relation (25) is approximated as

$$\omega^2 = k^2 \left(1 + 4\frac{\nu^2}{\omega^2} - \mu^2 \frac{h_1}{h_2} \frac{k^2}{3} \right), \tag{28}$$

and restricting the analysis to left-going waves only, one has

$$\omega k = -k^2 - 2\nu^2 + \mu^2 \frac{h_1}{h_2} \frac{k^4}{6}. \tag{29}$$

This expression applies to straight-crested waves propagating along the x-direction; to include a slow variation in the transverse (z-) direction, we write

$$k = (k_x^2 + \mu^2 k_z^2)^{1/2} = k_x + \frac{1}{2}\mu^2 (k_z^2/k_x) + O(\mu^4), \tag{30}$$

so, to the same order of approximation (29) becomes

$$\omega k_x = -k_x^2 - \frac{1}{2}\mu^2 k_z^2 - 2\nu^2 + \mu^2 \frac{h_1}{h_2} \frac{k_x^4}{6}. \tag{31}$$

It is now clear that rotation is as important as dispersive and transverse-variation effects if $\nu = O(\mu)$, as stated earlier in (24). Assuming this balance, by inspection of (31), the linear evolution equation for the interface elevation $\eta(x, z, t)$ is found to be

$$\eta_{Tx} = \eta_{xx} + \mu^2 \left\{ \frac{1}{6}\frac{h_1}{h_2}\eta_{xxxx} + \frac{1}{2}\eta_{zz} - 2\beta^2\eta \right\}. \tag{32}$$

To derive the complete evolution equation, we need to incorporate weakly nonlinear effects into (32). The appropriate nonlinear term is most conveniently obtained from the familiar KdV equation that is valid in the same flow configuration for straight-crested ($\partial/\partial z = 0$) long waves in the absence of rotation ($\beta = 0$):

$$\eta_t = \eta_x + \frac{3}{2}\epsilon \left(1 - \frac{h_2}{h_1} \right) \eta\eta_x + \frac{1}{6}\mu^2 \frac{h_1}{h_2}\eta_{xxx}, \tag{33}$$

where ϵ stands for the nonlinear parameter. Upon comparison of (32) with (33), the full rotation-modified KP equation in a reference frame moving with the long-wave speed reads

$$\eta_{Tx} - \frac{3}{4}\left(1 - \frac{h_2}{h_1} \right)(\eta^2)_{xx} - \frac{1}{6}\frac{h_1}{h_2}\eta_{xxxx} - \frac{1}{2}\eta_{zz} + 2\beta^2\eta = 0, \tag{34}$$

where $\epsilon = \mu^2$ and $T = \mu^2 t$ is a slow time. In addition, at the channel side walls, located at $z = 0, W$, the transverse velocity should vanish, which leads to the boundary conditions

$$\eta_z + 2\beta\eta = 0 \quad (z = 0, W). \tag{35}$$

The choice of initial conditions for the evolution equation (34) is not entirely straightforward and warrants clarification. As noted by Grimshaw (1985), locally confined solutions of (34), subject to the boundary conditions (35), must satisfy the integral constraint

$$\int_{-\infty}^{\infty} \eta(x, z, T)\, dx = M(T)e^{-2\beta z}, \tag{36}$$

where $M(T)$ is a function of T only. This condition apparently imposes a rather severe restriction on initial conditions, as the wave profile has to be such that the mass in planes parallel to the channel walls varies exponentially as $\exp(-2\beta z)$ across the channel. This requirement may raise doubts about the validity of the rotation-modified KP equation when $\eta(x, z, T = 0)$ does not observe (36) and, moreover, suggests that solitary wave solutions are unlikely to exist. In fact, Grimshaw(1985) was not successful in his attempt to find such solutions in the form suggested by the experiments.

The origin of this apparent difficulty can be traced to the approximate linear dispersion relation (31), according to which the group velocity along the channel, $\partial\omega/\partial k_x \to \infty$ as $k_x \to 0$, unless

$$k_z^2 + 4\beta^2 = 0. \tag{37}$$

Condition (37), along with (35), is entirely equivalent to the integral constraint (36). This reveals that (36) is directly related to the behavior of wave components with $k_x \ll 1$ and ensures that the group velocity remains finite so that the disturbance is localized for $T > 0$; in case the initial condition does not observe (36), the wave disturbance cannot be expected to remain locally confined, but would rather extend to $x = \infty$ for any $T > 0$, as verified by Ablowitz and Wang (1997) by singular perturbation analysis for small times.

It is worth noting that wave components with $k_x \ll 1$ vary more slowly along x than along z, so the approximation (30) of weak transverse variations breaks down, and hence they are not described properly by the rotation-modified KP equation (34). This issue was addressed by Grimshaw and Melville (1989) by solving the linear initial-value problem for the full governing equations; it turns out that the roration-modified KP equation does provide the correct asymptotic description of the far-field response, but the appropriate initial condition, obtained via matching with the inner field, is not locally confined in general owing to the radiation of Poincaré waves behind the main disturbance.

The rotation-modified KP equation (34), subject to the boundary conditions (35), was solved numerically by Katsis and Akylas (1987), using as initial condition a straight-crested KdV solitary wave with amplitude decaying exponentially as $\exp(-2\beta z)$ across the channel. Figure 1a and b shows the wave profile at $T = 2.5$ for two values of β corresponding to two of the various angular speeds used in

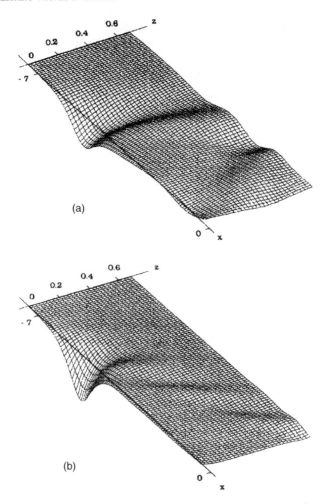

Figure 1: Three-dimensional wave profile at $T = 2.5$. (a) Low rotation, $\beta = 1.62$; (b) high rotation, $\beta = 4.83$ (from Katsis and Akylas 1987).

the experiments of Renouard *et al.* (1987). Consistent with the observations, the amplitude of the leading disturbance decays across the channel and the wave crest is curved backwards; moreover, for both values of the rotation parameter β, the main disturbance is accompanied by a train of smaller-amplitude waves that are left behind. As expected, this is not a wave of permanent form; the amplitude decays slowly as the disturbance propagates along the channel owing to the waves shed behind.

In further development of the theory, Melville *et al.* (1989) studied in more detail the mechanism that causes wave-crest curvature, drawing attention to the role of Poincaré waves − the small-amplitude waves trailing the main disturbance. Through an approximate analysis, valid for small times, Melville *et al.* (1989)

showed that an initially straight-crested KdV solitary wave with exponential decay across the channel, like the one used as initial condition in the numerical study of Katsis and Akylas (1987), generates Poincaré waves, which when superposed on the initial wave give rise to an apparent wave-front curvature. The same conclusion was reached by Grimshaw and Tang (1990) who also showed, using a small-time asymptotic analysis of the rotation-modified KP equation, that the generation of Poincaré waves results in radiation damping. Finally, Akylas (1991) conducted an asymptotic study of (34) and (35) in the weak-rotation limit ($\beta \ll 1$) and derived a nonlinear evolution equation that describes the decay of the amplitude of a solitary wave owing to radiation damping.

Assuming that transverse variations are relatively insignificant, the rotation-modified KP equation (34) reduces to

$$\left\{ \eta_T - \frac{3}{2}\left(1 - \frac{h_2}{h_1}\right)\eta\eta_x - \frac{1}{6}\frac{h_1}{h_2}\eta_{xxx} \right\}_x + 2\beta^2\eta = 0. \tag{38}$$

This evolution equation was first derived by Ostrovsky (1978) to account for the effects of the earth's rotation on ocean internal solitary waves and has been the subject of several studies (see Grimshaw et al. 1998b). Locally confined solutions of the Ostrovsky equation (38) clearly must satisfy the 'zero-mass' constraint:

$$\int_{-\infty}^{\infty} \eta \, dx = 0. \tag{39}$$

Similar to the constraint (36) of the rotation-modified KP equation discussed earlier, (39) is closely related to the dispersion relation of (38) being such that the group velocity $d\omega/dk \to \infty$ for very long waves ($k \to 0$). As a result, if the initial condition does not observe (39), a long-wave disturbance is emitted infinitely fast, adjusting the mass so that (39) is met for any $T > 0$ (Grimshaw 1999).

A notable feature of the Ostrovsky equation (38) is that it does not accept solitary wave solutions, and it is natural to ask what happens to a KdV solitary wave when the effect of rotation ($\beta \neq 0$) is taken into account. This issue was examined by Grimshaw et al. (1998a) using asymptotic methods for $\beta \ll 1$, as well as numerically. It turns out that the wave amplitude decays owing to the radiation of a dispersive tail, and the asymptotic theory suggests terminal damping of the solitary wave, as the amplitude is found to vanish at a finite time. The numerical results confirm this scenario, but also reveal that a new smaller-amplitude solitary wave emerges from the radiation, which again is damped, and the process keeps repeating itself. In the ocean, the radiation damping of internal solitary waves due to the earth's rotation appears to be comparable in strength to viscous damping.

Acknowledgments

The preparation of this chapter was supported in part by the Air Force Office of Scientific Research, Air Force Materials Command, USAF, under Grant FA 9950-04-1-0125, and by the National Science Foundation, under Grant DMS-0305940.

References

Ablowitz, M.J. & Wang, X.-P., Initial time layers and Kadomtsev–Petviashvili-type equations. *Stud. Appl. Math.*, **98**, pp. 121–137, 1997.

Akylas, T.R., On the excitation of long nonlinear water waves by a moving pressure distribution. *J. Fluid Mech.*, **141**, pp. 455–466, 1984.

Akylas, T.R., On the radation damping of a solitary wave in a rotating channel. *Mathematical Approaches in Hydrodynamics*, ed. T. Miloh, pp. 175–181, SIAM: Philadelphia, 1991.

Akylas, T.R. & Grimshaw, R.H.J., Solitary internal waves with oscillatory tails. *J. Fluid Mech.*, **242**, pp. 279–298, 1992.

Benney, D.J., Long nonlinear waves in fluid flows. *J. Math Phys.*, **45**, pp. 52–63, 1966.

Benney, D.J. & Ko, D.R.S., The propagation of long large-amplitude internal waves. *Stud. Appl. Math.*, **59**, pp. 187–199, 1978.

Derzho, O. & Grimshaw, R., Solitary waves with recirculation zones in axisymmetric rotating flows. *J. Fluid Mech.*, **464**, pp. 217–250, 2002.

Grimshaw, R., Evolution equations for weakly nonlinear long internal waves in a rotating fluid. *Stud. Appl. Math.*, **73**, pp. 1–33, 1985.

Grimshaw, R., Resonant flow of a rotating fluid past an obstacle: the general case. *Stud. Appl. Math.*, **83**, pp. 249–269, 1990.

Grimshaw, R.H.J., Adjustment processes and radiating solitary waves in a regularized Ostrovsky equation. *Eur. J. Mech. B/Fluids*, **18**, pp. 535–543, 1999.

Grimshaw, R. & Melville, W.K., On the derivation of the modified Kadomtsev–Petviashvili equation. *Stud. Appl. Math.*, **80**, pp. 183–202, 1989.

Grimshaw, R. & Smyth, N., Resonant flow of a stratified fluid over topography. *J. Fluid Mech.*, **169**, pp. 429–464, 1986.

Grimshaw, R. & Tang, S., The rotation-modified Kadomtsev–Petviashvili equation: an analytical and numerical study. *Stud. Appl. Math.*, **83**, pp. 223–248, 1990.

Grimshaw, R. & Yi, Z., Resonant generation of finite-amplitude waves by the flow of a uniformly stratified fluid over topography. *J. Fluid Mech.*, **229**, pp. 603–628, 1991.

Grimshaw, R.H.J., He, J.-M. & Ostrovsky, L.A., Terminal damping of a solitary wave due to radiation in rotational systems. *Stud. Appl. Math.*, **101**, pp. 197–210, 1998a.

Grimshaw, R.H.J., Ostrovsky, L.A., Shrira, V.I. & Stepanyants, Yu. A., Long nonlinear surface and internal gravity waves in a rotating ocean. *Nonlinear Processes in Geophysics*, **19**, pp. 289–338, 1998b.

Katsis, C. & Akylas, T.R., Solitary internal waves in a rotating channel: a numerical study. *Phys. Fluids*, **30**, pp. 297–301, 1987.

Leibovich, S., Weakly nonlinear waves in rotating fluids. *J. Fluid Mech.*, **42**, pp. 803–822, 1970.

Leibovich, S., Vortex breakdown: a coherent transition trigger in concentrated vortices. *Turbulence and Coherent Structures*, eds. O. Metais & M. Lesieur, pp. 285–302, Kluwer: Dordecht, 1991.

Maxworthy, T., Experiments on internal solitary waves. *J. Fluid Mech.*, **129**, pp. 365–383, 1983.

Melville, W.K., Tomasson, G.G. & Renouard, D.P., On the stability of Kelvin waves. *J. Fluid Mech.*, **206**, pp. 1–23, 1989.

Ostrovsky, L.A., Nonlinear internal waves in a rotating ocean. *Oceanology*, **18**, pp. 119–125, 1978.

Renouard, D.P., Chabert d'Hières, G. & Zhang, X., An experimental study of strongly nonlinear waves in a rotating system. *J. Fluid Mech.*, **177**, pp. 381–394, 1987.

Veronis, G., The analogy between rotating and stratified fluids. *Annu. Rev. Fluid Mech.*, **2**, pp. 37–66, 1970.

Yih, C.-S., *Fluid Mechanics*, West River Press: AnnArbor, 1979.

CHAPTER 6

Planetary solitary waves

J.P. Boyd
University of Michigan, Ann Arbor, MI, USA.

Abstract

This chapter describes solitary waves in the atmosphere and ocean with diameters from perhaps a hundred kilometers to scales larger than the earth: vortices embedded in a shear zone, Rossby solitons and equatorial Kelvin solitary waves among others. One theme is that while nonlinear coherent structures are readily identified in both observations and numerical models, it is difficult to determine when these have the balance of dispersion and nonlinear frontogenesis that is central to classical soliton theory, and also unclear whether the quest for this balance is illuminating. A second theme is that periodic trains of vortices should be regarded as solitary waves, but usually are not, because the vortices are dynamically isolated even though they are geographically close. A third theme is the generalization to radiating or weakly nonlocal solitary waves: all solitons slowly decay by viscosity (in nature if not always in theory), so it is foolish to exclude coherent structures that have in addition an even slower decay through radiation. No observation of a large nonlinear coherent structure has been universally accepted as a soliton large enough in scale to feel Coriolis forces. The reasons may have more to do with overly narrow concepts of a 'solitary wave' than to a lack of data. The Great Red Spot of Jupiter and Legeckis eddies in the terrestrial equatorial ocean are well-known vortices for which an appropriately broad notion of solitons is probably useful. In contrast, hurricanes, although nonlinear coherent structures, are so strongly forced and damped that it is these mechanisms and also the complex feedback from the cloud scale to the vortex scale and back that are rightly the focus, and not the ideas of soliton theory.

1 Introduction

Persistent, spatially localized structures are a ubiquitous feature of the ocean and atmosphere and of the atmospheres of other planets. Vortices with diameters of a couple of thousand kilometers or so are the heart of large-scale weather patterns in the midlatitudes. Gulf Stream rings are intense ocean vortices, lasting for many months, or even years, that are spawned by the Gulf Stream, the oceanic equivalent of hurricanes. The Great Red Spot of Jupiter is an anticyclonic vortex that has been spinning for at least three centuries, so vast that Earth would fit comfortably within it. Long-lived, spatially localized structures of large scale or planetary scale are clearly of great importance.

Consequently, so much has been written about Rossby vortices and coherent structures, as catalogued in the books and reviews of Table 1, that it is impossible to give a complete treatment in a brief space. Indeed, an entire monograph has

Table 1: Selected reviews, monographs and special issues on Rossby solitary waves and coherent vortices.

Reference	Topics, always including Rossby solitons and vortices
Miles (1980)	Solitary waves including Rossby solitons
Malanotte Rizzoli (1982)	Rossby solitons in channels
Ho and Huerre (1984)	Perturbed free shear layers
McWilliams (1985)	Small-scale (submesoscale) ocean vortices
Maslowe (1986)	Critical layers in shear flows
Flierl (1987)	Ocean vortices
Boyd (1989a)	Polycnoidal waves; Rossby solitons
Nihoul and Jamart (1989)	Conference proceedings on ocean vortices
Hopfinger and van Heijst (1991)	Laboratory experiments on solitons and modons
Boyd (1991a)	Nonlinear equatorial waves
Boyd (1991b)	Weakly nonlocal solitary waves
Boyd and Haupt (1991)	Polycnoidal waves
Olson (1991)	Observations of ocean vortices
Sommeria et al. (1991)	Laboratory experiments
Meleshko and van Heijst (1992)	Chaplygin and Lamb's discovery of modons
Nezlin and Snezhkin (1993)	Monograph on Rossby vortices and solitons
Nezlin and Sutyrin (1994)	Vortices on gas giant planets
Marcus (1993)	Great Red Spot of Jupiter
Chaos, vol. 4 (1994)	Special issue on coherent structures
Korotaev (1997)	Weakly nonlocal Rossby vortex solitons
Boyd (1998b)	Nonlocal, radiating solitary waves – one chapter on Rossby solitons
Ingersoll (1999, 2002)	Good images of outer planet vortices
Barcilon and Drazin (2001)	Nonlinear vorticity waves
Carton (2001)	Ocean vortices
Vasavada and Showman (2005)	Jovian dynamics; Great Red Spot

already been devoted to the subject (Nezlin and Snezhkin 1993). There are literally hundreds of interesting articles that cannnot be discussed here because of space limitations. Instead, we will concentrate on a handful of major themes, important both to solitary wave theory in general and to Rossby solitons in particular.

In a long review focused entirely on Rossby solitons, Paola Rizzoli wrote in 1982:

> . . . unambiguous experimental evidence has not yet been produced for the existence of solitary Rossby waves, either in the ocean or in the atmosphere.

Nearly a quarter of a century later, this is still true: there are no universally-accepted observations of Rossby solitary waves. There are, however, many promising candidates as discussed below. In part, the lack of accepted soliton sightings is because the historical development of 'soliton theory' has led to definitions and mental models that are too narrow. In the next three sections, we elaborate on this theme.

2 Coherent structures and solitons

Solitary waves are a subset of a much larger family of 'coherent structures'. The formal definitions of each are as follows:

Definition 1 (Coherent structure): *a steadily translating, finite-amplitude, spatially localized disturbance that does not evolve in time, or evolves only slowly.*

Definition 2 (Solitary wave): *a solitary wave, also known as a 'soliton', is a coherent structure that does not evolve in time because of a perfect balance between the steepening effects of nonlinearity and the spreading effects of wave dispersion.*

The crucial distinction is that in the classical theories of solitary waves, such as those of the Korteweg–de Vries equation, the longevity and persistence of amplitude, speed and shape are attributed to a *specific mechanism*: the balance between nonlinearity and dispersion. The complication is that vortices and other coherent structures can have many of the characteristics of solitary waves without sharing this steepening-balances-dispersion mechanism.

Two-dimensional turbulence in an incompressible fluid without Coriolis forces is perhaps the clearest illustration. McWilliams (1984) showed that an initial state, which is filled with many small vortices of random amplitudes and phases will spontaneously evolve to a state with a few vortices, separated by large gulfs free of organized vortices, through repeated vortex mergers. Twenty years later, he observed (McWilliams 2004: p. 38):

> The most important change [in our understanding of turbulence] is the realization that the generic behaviour of turbulence is the development of coherent structures at high-Re values. The case for this was first and best made in 2D turbulence because of its computational affordability; however, this phenomenon has since been confirmed in many turbulent regimes, albeit with different typical flow patterns in different regimes.

Two-dimensional incompressible turbulence is an extreme illustration of coherent-structures-that-are-not-solitary-waves because the equations of motion are completely *wave-free*. There are no sound waves because the fluid is incompressible, no gravity waves because there are no buoyancy forces, and no Rossby waves because there are no Coriolis forces. Nevertheless, coherent structures – McWilliam's 'long-lived, isolated, axisymmetric monopole' vortices – spontaneously emerge as illustrated in Fig. 1.

Table 2 catalogues the many similarities between disc-shaped or elliptical-shaped vortices (in two dimensions) or cylindrical or elliptic-cylindrical vortices (in three dimensions), and solitary waves. Even more dramatically, hurricanes and typhoons have similarities to solitons: they are strongly nonlinear, emerge spontaneously from initial conditions very different from the final structure, are

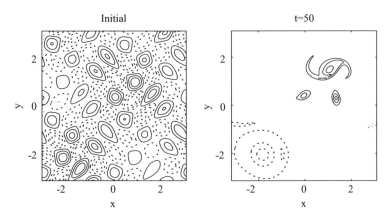

Figure 1: Contour plots of the vorticity for typical two-dimensional turbulence at the initial time (left) and at $t = 50$ (right). Positive-valued isolines are solid; negative contours are dashed.

Table 2: Properties of vortices and solitary waves.

Property	Solitary wave	Coherent vortices
Spatially localized	Yes	Yes
No time evolution except for steady translation and collisions and near-collisions	Yes	Yes
Form spontaneously from highly nonsolitonic or noncoherent structure initial conditions	Yes	Yes
Nonlinearity essential	Yes	Yes
Balance between nonlinearity and dispersion	Yes	Sometimes

spatially localized and are immortal in the absence of interactions. (For hurricanes, landfall is fatal because this chokes off the supply of moist surface air that sustains the storm; on an 'aqua planet', free of land, a hurricane could spin and roar forever.) Why aren't hurricanes classed as solitary waves? How can Rizzoli's claim of 'unambiguous experimental evidence has not yet been produced' be made without laughing?

The answer is partly pragmatism and partly history. The pragmatism is that although there indeed have been a few papers that have identified hurricanes as solitons, or at least linked the storm with the concept (Zhao *et al.* 2001), a hurricane is both *strongly forced* and *strongly dissipative*. The latent heat forcing is what makes hurricanes so powerful; the equally strong damping is why hurricanes die quickly after landfall when the forcing is choked off. The mathematical folklore known as 'soliton theory', pregnant with one-space-dimensional unforced and undamped equations like the Korteweg–de Vries equation, is simply not useful to modern hurricane specialists. A typhoon modeler's dreams are haunted by the vast separation of scales between the parent vortex and the individual cumulus 'hot towers' that convert moist air into clouds, rain and terror. The triumphs of the KdV/NLS crowd are as remote and irrelevant as palm trees to a penguin.

The history is that the solitary wave theory exploded in the 1960s with the coining of 'soliton' and the development of the inverse scattering theory for solving a host of exactly integrable equations. In contrast, the subtlety that an axisymmetric or slightly elongated vortex may be very like a solitary wave, and yet have no wave dispersion at all, was rather slow emerging. McWilliams' quote is taken from a *festschrift* in honor of the 75th birthday of Doug Lilly, who invented what is now called LES turbulence modeling and did the first numerical studies of two-dimensional turbulence (Lilly 1969,1971). It continues

> ... In hindsight the characteristic long-lived isolated, axisymmmetric, monopole patterns in $\zeta(x, y)$ [vorticity] – coherent vortices – are not evident in Lilly's published figures. (He has subsequently remarked that he did see indications of them in evidence not published.)

Lilly's coarse 64×64 second order finite difference computations could only hint at the emergence of coherent, isolated structures, and Lilly did not think the hints worth publishing. Eight years later in 1977, Fornberg applying a more powerful pseudospectral method, observed

> the flow simplifies into a few 'cyclones' or 'finite area vortex regions'.

A comprehensive analysis of vortex formation had to wait another seven years, however, and the rise of vector supercomputers (McWilliams 1984).

Meanwhile, beginning with the MODE and POLYMODE field programs of the mid-70s, oceanographers had discovered that the sea is filled with a bewildering variety of vortices (McWilliams 1983,1991). At the turn of the 20th century, Chaplygin and Lamb had independently discovered exact solutions for vortex pairs in two-dimensional flow (Meleshko and van Heijst 1992). On the eve of MODE, Stern (1975) and Larichev and Reznik (1976) independently generalized these solutions to include the beta-effect, the westward wave propagation created by the latitudinal derivative of the Coriolis parameter, which is always denoted by the Greek letter β. Stern coined the catchy term 'modons' to describe such contra-rotating vortex pairs.

Although modons have been the subject of intensive research down to the present, modon theory has not been a huge success in explaining ocean vortices. The reason is that ocean vortices like Gulf Stream rings, meddies and so on are primarily *monopoles*, i.e. nearly axisymmetric discs or cylinders of a *single* sign of vorticity. Modons have also been invoked to explain atmospheric blocking, which is the formation of stationary, relatively long-lived transient features that deflect storm systems around them (McWilliams 1980; Haines and Marshall 1987; Malanotte Rizzoli and Hancock 1987; Flierl and Haines 1994). However, real atmospheric blocks are strongly modified by forcing, damping and topography.

Vortex pairs are common features in the ocean as eloquently documented with satellite images in (Fedorov and Ginsburg 1986), but most of the observed vortex pairs, and many of the monopoles, are too small in scale for Coriolis forces and the beta effect to be important. Like the disc-shaped eddies that emerge in unforced two-dimensional Coriolis-free turbulence, many ocean vortices must be understood as vortices and not as solitons. All too often, theorists have tried to shoehorn wave physics and dispersion into observed eddies so as to deploy the heavy artillery of soliton theory even when the target is a vortex but not a wave.

3 Weakly nonlocal solitary waves: slow death by radiation

The Korteweg–de Vries equation has a very simple linearized dispersion relation: infinitesimal waves proportional to $\exp(ikx)$ always have negative phase speeds relative to the long wave speed c_0, and $|c - c_0|$ increases monotonically with k. In contrast, the solitary waves always travel faster than c_0. In the 'far field' of the soliton where its amplitude has decayed to a very small value and the linear dispersion relation again applies, such faster-than-linear speeds require *complex-valued* k, which is a highbrow way of stating that the solitary wave *decays* exponentially fast as $|x| \to \infty$. This rapid asymptotic-in-space decay has seemed an essential property of a soliton, a necessary guarantor that the structure is truly 'solitary'.

However, many wave equations have complicated, nonmonotonic linear dispersion relations. The solitary wave may then be resonant with small-amplitude waves of some wavenumber k_f in the sense that the speed of the strongly nonlinear solitary wave matches that of an infinitesimal wave of wavenumber $k = k_f$. Travelling waves still exist, but the solitary waves are 'weakly nonlocal' in the sense that the core of the solitary wave does not asymptote to zero, but rather to a quasi-sinusoidal oscillation proportional to $\alpha \exp(ik_f[x - ct])$. If α, the 'radiation coefficient', is small compared to the amplitude of the core of the traveling wave, then the coherent structure is a 'weakly nonlocal solitary wave' (Fig. 2).

On an unbounded domain, these traveling waves have the absurd property of infinite energy. In reality, a spatially localized initial condition will spontaneously generate one or more soliton cores, and these will *radiate* waves of wavenumber k_f and amplitude α so that the core slowly decays by radiation.

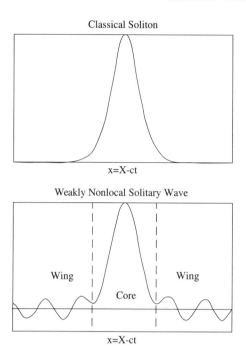

Figure 2: Comparison of a classical soliton (top) with a weakly nonlocal soliton (bottom).

The notion of a 'nonlocal' or 'radiating' solitary wave still meets with resistance. However, real fluids are always viscous, which implies that a classical, exponentially decaying-at-infinity solitary wave is only an approximation to a physical reality that is always slowly decaying in time. It seems silly to be selectively (and usefully) blind to viscous decay, if very slow, and not extend the same attitude towards radiative decay, if slower still.

The subtleties of nonlocal solitary waves in general are discussed in the author's book (Boyd 1998b) and reviews (Boyd 1989a, 1991b, 1999). One chapter of (Boyd 1998b) discusses weakly nonlocal Rossby solitary waves using drastically simplified models. The reviews by Flierl (1994) and Korotaev (1997) are focused on nonlocal Rossby vortices.

The relevance of all this to geophysics is that Rossby waves can be weakly nonlocal in at least four ways:

1. resonances with different vertical modes,
2. resonances because the soliton is travelling within the range of linear Rossby phase speeds,
3. radiation into equatorially trapped waves of lower latitudinal mode number,
4. resonances with other species of waves.

For example, weakly nonlinear solitary waves whose energy is concentrated in the first baroclinic vertical mode will be coupled by nonlinearity to all the other modes

(Marshall and Boyd 1987). For equatorially trapped Rossby waves, for example, the barotropic vertical mode is not equatorially trapped. Energy, which is transferred from the first baroclinic mode by nonlinearity to the barotropic mode, therefore radiates away from the equator to higher latitudes, thus providing a radiative damping for equatorially trapped baroclinic Rossby solitons. Figure 1 of Williams (1996) is a beautiful numerical illustration of this process. However, the rate of leakage is small so that the weakly nonlocal solitary wave is very long-lived. Flierl's (1994) review is a good description of similar barotropic radiation of a baroclinic midlatitude, nearly circular vortex.

However, a vortex will also radiate if its speed is in the range allowed for linear Rossby waves. Since a single vortex – a monopole – is motionless except for the beta effect, it is always in the linear range and always radiating.

If a modon is formed (in the northern hemisphere) from an anticyclone north of a cyclone, the mutual interaction of the two vortices will propel them westward. If the vortices are sufficiently strong, the westward speed will be larger than the maximum phase speed of linear Rossby waves, and therefore resonant coupling to linear waves in the far field is impossible, and the modon will be a classical, nonradiating coherent structure. However, a modon with weak vortices will travel westward so slowly that it will radiate Rossby waves. Flierl and Haines (1994) give a thorough analysis.

Linear equatorial Rossby waves have a modal structure in latitude which is described by Hermite functions, as will be explained in detail in Section 5 below. In the shallow-water model described there, which neglects coupling between different vertical modes, Rossby solitary waves in the lowest symmetric mode, the $n = 1$ latitudinal mode, are classical, nonradiating solitary waves. However, Williams and Wilson (1988) observed numerically that higher-mode solitary waves radiated Rossby waves of lower latitudinal wavenumber, which are more tightly confined to the vicinity of the equator. This is really a subcase of the previous case: the $n = 3$ solitary wave has a nondimensional phase speed of about $-1/7$, but linear waves in the $n = 1$ latitudinal mode can travel as fast as $c = -1/3$: the higher-mode equatorial solitons travel within the range of linear Rossby wave phase speeds, and consequently radiate. Boyd (1989b, 1998b) shows that the coupling is *exponentially* small in the reciprocal of the amplitude of the core of the higher-mode solitary waves, so *small-amplitude* higher-mode solitary waves are extremely long-lived.

Surface gravity water waves without Coriolis forces were the first class of solitons studied: everything in solitary wave theory is the child of the pioneering 19th century studies of Russell, Boussinesq, Rayleigh and Korteweg and deVries. Ironically, this prototype of a solitary wave is actually a weakly nonlocal solitary wave: the soliton is resonantly coupled with very short capillary waves. However, the amplitude of the capillary radiation is exponentially small in the relevant parameters, so it is only recently that the existence of this resonant coupling has been recognized (Boyd 1998b). Later, we will discuss resonances between equatorial Kelvin wave solitons and eastward-traveling gravity waves. Indeed, there are so many different vertical modes/latitudinal modes/wave species to resonate with that classical solitary waves, for all but the most highly idealized systems, are more the rule than the exception (Medvedev and Zeitlin 2005; Reznik and Zeitlin 2006).

Some wave species may be either classical solitary waves or weakly nonlocal solitary waves. Westward-propagating modons on a beta-plane or on the sphere are classical solitary waves if their amplitude is sufficiently *large* so that the nonlinear interaction of the two vortices accelerates them westward faster than the fastest linear Rossby wave (Boyd 1994). However, a *weak* westward modon will resonantly leak small-amplitude Rossby waves to infinity and is only a weakly nonlocal coherent structure. Indeed, numerical studies of the initial value problem with viscosity have shown that a strong modon may decay very slowly, weakening only because of viscosity, for a long time until its speed slows to that of the fastest linear wave, and then the decay rate dramatically accelerates as radiative damping is added to viscous damping. Tribbia (1984) and Verkley (1984, 1987, 1990) have calculated modons on the sphere that consist of a moderately strong vortex pair accompanied by a 'dressing' of small-amplitude, quasi-sinusoidal Rossby waves that fill the rest of the globe.

4 The dynamical isolation of the crests of cnoidal waves: Gauss' cosine-lemniscate function, imbricate series and all that

The label 'solitary wave' conveys a mental image of a single disturbance, alone on a wide, wide sea. But this image is misleading. Most types of solitary waves decay *exponentially* fast with distance from the maximum of the soliton. It follows that two solitary waves can be rather close together, and yet nevertheless be *dynamically* isolated in the sense that their overlap is exponentially weak and their interaction therefore exponentially small. The image conveyed by the title of a famous sociological treatise, *The Lonely Crowd*, which described multitudes of noninteracting persons packed into cities, not talking, is sometimes a more sensible model of solitary waves.

For the Korteweg–de Vries equation, it is possible to be precise about the physics of a 'lonely crowd'. Korteweg and de Vries showed that the equation that bears their name has traveling waves that they dubbed 'cnoidal waves' because these solutions could be expressed in terms of the elliptic cosine function, whose mathematical symbol is 'cn'. For a given spatial period, the cnoidal waves depend on amplitude, like a soliton, but also on an additional parameter m, the elliptic modulus. When $m = 0$, cnoidal waves are cosine waves of infinitesimal amplitude. When $m = 1$, the peaks are infinitely tall and narrow and have the $sech^2$ shape of solitary waves.

More than three-quarters of a century later, Morikazu Toda, best known for the 'Toda' lattice, discovered the remarkable fact that the cnoidal wave is given for all values of the elliptic modulus as the superposition of solitary waves (Toda 1975):

$$u_{\text{cnoidal}}(X) = 2\pi M - 12\frac{\epsilon}{\pi} + 12\epsilon^2 \sum_{j=-\infty}^{\infty} sech^2 \left(\epsilon(X - 2\pi j) \right), \qquad (1)$$

where M, an arbitrary constant, is the mean of the wave on the period, $X \equiv x - ct$, and the corresponding phase speed is $c = 2\pi M - 12\frac{\epsilon}{\pi} + 4\epsilon^2 - 24\epsilon^2 \sum_{n=1}^{\infty} 1/\sin^2(n\, 2\, \pi\, \epsilon)$. Note that although the shape of the solitons is not changed by their interaction, the phase speed is altered. The series of solitons, which is the 'imbrication' of the soliton, converges very rapidly when $\epsilon \gg 1$ so that the peaks are tall and narrow and have very little overlap. Remarkably, the imbricate series still converges, albeit not rapidly, even when ϵ is small and the cnoidal wave is very like a cosine.

In the sense that the imbricate series and Fourier series converge at equal rates, the maximum overlap between the cosine wave and soliton regimes is at $m = 1/2$. The cnoidal wave $-0.95493 + 2.08975\,\mathrm{cn}^2(0.5902x; m = 1/2)$ can be written as the Fourier series,

$$u(X; m = 1/2) = 1.0391 \cos(x) + 0.0448 \cos(2x) + 0.0019 \cos(3x) + \cdots, \quad (2)$$

which shows the very rapid convergence of the Fourier series. For this special case, the zero-mean cnoidal wave has an exact phase speed of -0.95493. The phase speed of an infinitesimal amplitude cosine wave, $c_{\cos} = -1$, is in error by only 4.7%, consistent with the fact that Fourier series is dominated by just one term. The phase speed of a solitary wave (with correction so that the mean of the imbricate series is zero) is $c_{\mathrm{sol}} = 1 - 6/\pi = -0.90986$, which is in error by the same 4.7% in the opposite direction. The KdV cnoidal wave for this value of the elliptic modulus is equally both cosine and soliton.

Gauss studied this special case in 1797 and defined the 'cosine-lemniscate' function as the inverse of an integral that arose in computing arcs of a lemniscate (Whittaker and Watson 1940: p. 524):

$$x \equiv \mathrm{coslemn}(\phi) = \mathrm{cn}(\sqrt{2}\phi; m = 1/2) \quad \leftrightarrow \quad \phi = \int_x^1 \frac{1}{\sqrt{1 - t^4}}\, dt. \quad (3)$$

Two centuries later, there is another reason to be intrigued by $m = 1/2$. The cosine-lemniscate case shows, as bluntly as a club between the eyes, that the boundary between periodic wave and soliton is nonexistent.

As a further example, the traveling waves of the nonlinear Schrödinger (NLS) equation and the modified-Korteweg–de Vries (MKdV) equation both solve $u_{xx} - \lambda u + 2u^3$. The cosine-lemniscate is a solution for $\lambda = 0$, and the NLS/MKdV cnoidal wave can again be expressed equally well as a Fourier series or as a periodized series of solitons – this time, with alternating signs:

$$u = 0.8346\,\mathrm{coslemn}\,(0.8346\,x) \qquad (4)$$

$$= \sum_{j=-\infty}^{\infty} (-1)^j \,\mathrm{sech}\,(x - j\pi)) \qquad (5)$$

$$= 2 \sum_{n=1}^{\infty} \mathrm{sech}\left(\frac{(2n-1)\pi}{2}\right) \cos((2n-1)x). \qquad (6)$$

Figure 3 shows how closely the cosine and soliton approximations agree with the cosine-lemniscate function.

Kindle's (1983) investigation is a very good numerical study of equatorial Rossby solitary waves, which are described in the next section. He showed that a change

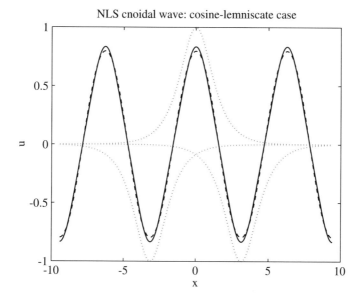

Figure 3: Thick solid curve: lemniscate case of the cnoidal wave of the NLS equation, which solves $u_{xx} + u^3 = 0$. Dashed curve: the cosine approximation $u \approx 2\text{sech}(\pi/2)\cos(x)$. Dotted: Three solitons, one positive and two negative, which are terms of the alternating imbricate series for the cosine-lemniscate function.

in wind stress acting on the tropical ocean would readily excite a train of solitary waves rather than just one, as confirmed in (Bouchut *et al.* 2005). However, a longer version of his article, submitted to the *Journal of Physical Oceanography*, was never published: reviewers simply could not accept the concept of a periodic sequence of solitary waves, or understand that geographic isolation is not required for dynamic isolation.

This point is important because solitary Rossby waves are as likely to occur in quasi-periodic arrays ('lonely crowds') as in splendid, utter isolation.

5 An exemplary model: equatorial solitary waves in the shallow-water equations

5.1 Background

The equatorial ocean has very rich dynamics. At the same time, it is comparatively easy to apply the method of multiple scales perturbation theory to illustrate by example how the Kortweg–deVries equation and other one-space-dimensional equations can be derived as sensible, systematic models for three-dimensional reality. We shall therefore discuss the various kinds of planetary solitary waves for this case in some detail.

The tropical ocean consists of a warm, well-mixed layer with an average thickness of about 100 m on top of cold abyssal waters with an average thickness 40 times greater. The mean temperature is almost a step function with the jump located at the 'main thermocline'. It is therefore not a half-bad approximation to pretend that the equatorial ocean is composed of two layers of homogeneous fluid of different densities. Because the lower layer is so deep compared to the upper layer, it is only a modest approximation to pretend that the lower layer is infinitely deep and therefore *motionless*. The dynamics of undulations on the thermocline is then described by the equations for a *single* layer of fluid with the actual depth replaced by the so-called 'equivalent' depth, which is the depth multiplied by the fractional density difference between the two layers. (The equivalent depth is only 0.4–0.6 m!) Since our interest is in planetary-scale motion, the hydrostatic approximation is a good approximation, and the equations reduce to the so-called 'shallow water' equations. Because our interest is in equatorially trapped waves, it is a good approximation to replace the sines and cosines of latitude by a linear function of latitude and by the constant one, respectively. All these approximations yield the very widely used model, which is called the 'nonlinear shallow-water wave equations on the equatorial beta-plane':

$$u_t + uu_x + vu_y - yv + h_x = 0, \tag{7}$$

$$v_t + uv_x + vv_y + yu + h_y = 0, \tag{8}$$

$$h_t + \{uh\}_x + \{vh\}_y = 0, \tag{9}$$

where u and v are the east–west and north–south velocities and h is the total depth of the fluid. The equations have been non-dimensionalized in the way standard in equatorial dynamics (Boyd 2002b). A unit length scale is about 300 km or three degrees of longitude and the velocity scale is about 2 m/s.

The weakly nonlinear theory of equatorial solitary waves has been developed by Boyd (1980b, 1983, 1984, 1985, 1989b, 1991a, 2002b), Marshall and Boyd (1987), Kindle (1983), Wu (1986), Williams and Wilson (1988), Long and Chang (1990), Greatbatch (1985), Jain *et al.* (1981), Zhao *et al.* (2001a), Le Sommer *et al.* (2004) and Bouchut, *et al.* (2005). The fundamental idea is the same as for any weakly nonlinear theory: the starting point is the theory of *linear* waves. In three dimensions, the linear solutions have the *separable* form

$$u(x, y, z, t) \approx A(x, t) Y_n(y) Z_m(z), \tag{10}$$

where n and m denote the eigenmode numbers for the functions, both solutions to one-dimensional eigenproblems, that give the latitudinal and vertical structure. (In the shallow-water model, the single vertical structure function $Z(z)$ is a step function, equal to one in the upper layer and zero below the thermocline; the three-dimensional equatorial case is treated in (Marshall and Boyd 1987.) In both latitude and depth, equatorial waves have the structure of *standing* waves; the wave is confined and does not propagate in those directions. In z, the barriers are the top and bottom of the ocean. In latitude, the barrier is invisible but no less real: the refractive effects of the variation of the Coriolis parameter with latitude create an *equatorial waveguide* so that the wave species of interest are confined to within a few degrees of the equator. Weakly nonlinear corrections alter the latitudinal and vertical structure only at higher order, and not in particularly interesting ways.

The longitude and time factor $A(x, t)$, however, can be drastically modified by nonlinearity even when the nonlinearity is weak. The reason is that the wave freely propagates in longitude, and as it propagates, small changes can *accumulate* so as to produce $O(1)$ alterations in the longitudinal shape of the wave. In particular, nonlinearity can counterbalance dispersion so that $A(x, t)$ will not disperse into an ever-wider, ever-smaller field of ripples as $t \to \infty$, but instead evolve into a solitary wave.

The usual KdV theory is logically a perturbation series in two independent parameters: the amplitude ϵ and $\delta = 1/L$, where L is the east–west spatial scale. However, when $\epsilon \gg \delta^2$, the nonlinearity dominates and the leading (westward) edge of the wave will steepen until the east–west structure of the flow has a near discontinuity. If $\epsilon \ll \delta^2$, dispersion will overwhelm nonlinearity and the wave will evolve as in linear wave theory. The only interesting limit is when $\epsilon \sim \delta^2$. Assuming this balance then reduces the perturbation theory to an expansion in a *single* parameter ϵ.

However, the nonlinear and dispersive effects remain physically and logically separate. Boyd gave an amusing demonstration of this logical separation by calculating the dispersive effects on Kelvin solitary waves, discussed later in the next section, four years after computing the nonlinear terms (Boyd 1980a, 1984)!

The amplitude expansion by itself would yield a series with a finite radius of convergence in ϵ. (For the amplitude expansion for the KdV cnoidal wave, for example, one can easily compute the convergence domain). The long-wave expansion has a much more drastic effect in that it, like most multiple scales expansions, is an asymptotic-but-divergent series with a *zero* radius of convergence (Boyd 1998b, 1999, 2005: pp. 553–575).

This would seem to make it desirable to make only an amplitude expansion, and forgo the long wave expansion. However, the latitudinal structure of infinitesimal equatorial Rossby waves varies with the zonal wavenumber k. The only way to obtain a *separable* structure, such that the lowest order approximation is the product of a factor of longitude and time only with a factor of latitude only, is to make the long-wave approximation, and thereby approximate the latitudinal structure for all k by its limit as $k \to 0$.

5.2 Derivation of KdV by multiple scales/small amplitude perturbation theory

The analysis below closely follows that of Boyd (1980b) (for equatorial waves) and Clarke (1971) for midlatitude Rossby waves in a channel. Good references on the method of multiple scales include (Nayfeh 1973; Bender and Orszag 1978). The equations to be solved are the nonlinear shallow-water wave equations on the equatorial beta-plane (7–9).

The phase speed of solitary waves can be split into two components: the linear, nondispersive phase velocity for the mode in question plus the (small) correction due to the perturbations owing to finite amplitude and finite zonal scale. It is convenient to shift into a coordinate system moving with the linear, nondispersive phase velocity for the chosen wave mode so that all time variations will be due to the perturbing effects of nonlinearity and dispersion. Define

$$s = x - c_0 t, \quad c_0 = -1/(2n + 1), \tag{11}$$

where n is the latitudinal mode number defined below. In the method of multiple scales, we normally define both 'fast' and 'slow' variables. Because of the change of coordinate (11), however, we have no 'fast' variables, but only the 'slow' variables

$$\xi = \epsilon^{1/2}(x - c_0 t) \quad \text{and} \quad \tau = \epsilon^{3/2} t. \tag{12}$$

The unknowns are expanded as

$$u = \epsilon\, u^0(\xi, y, \tau), \tag{13}$$

$$v = \epsilon^{3/2}\, v^0(\xi, y, \tau), \tag{14}$$

$$\phi = \epsilon\, \phi^0(\xi, y, \tau). \tag{15}$$

The formal way of deriving the exponents of ϵ in (12)–(15) is to allow each unknown to have a symbolic exponent, substitute in (32), and then match powers. The conditions that (1) the lowest order set of equations be the linear, 'long wave' or 'meridional geostrophy' approximation and (2) the dispersive and nonlinear terms all appear at first order then force us to choose the exponents as given above.

The physical explanation of the powers of ϵ in the new coordinates is that we chose the 'slow' zonal variable ξ so that a unit length scale in ξ corresponds to a scale of $L \sim O(1/\epsilon^{1/2})$, which in turn implies a dispersion that is $O(\epsilon)$. The scaling of τ implies that the phase speed will be altered by an amount of $O(\epsilon)$ by the nonlinear and dispersive correction; the extra $\epsilon^{1/2}$ is needed because the phase speed correction must be multiplied by the inverse of the zonal length scale to convert it to a frequency correction. Finally, the reason for the extra factor of $\epsilon^{1/2}$ multiplying v^0 is that the meridional velocity for an ultra-long Rossby wave is $O(1/L)$ smaller than u and ϕ. The lowest order set, the unperturbed equations, are

$$-c_0 u_\xi^0 - y v^0 + \phi_\xi^0 = 0, \tag{16}$$

$$y u^0 + \phi_y^0 = 0, \tag{17}$$

$$-c_0 \phi_\xi^0 + u_\xi^0 + v_y^0 = 0, \tag{18}$$

which are the *linear, long-wave* approximations. The only change from the full linear equations is the neglect of the time derivative in the north–south momentum equation, but this one omission is sufficient to (1) filter out all gravity waves and the mixed Rossby-gravity wave and (2) make Rossby waves nondispersive. The solutions are (Boyd 1980b)

$$v^0 = A_\xi(\xi, \tau)\psi_n(y), \tag{19}$$

$$u^0 = A(\xi, \tau)\left\{\frac{1}{1-c}\left[\frac{n+1}{2}\right]^{1/2}\psi_{n+1}(y) + \frac{1}{1+c}\left[\frac{n}{2}\right]^{1/2}\psi_{n-1}(y)\right\}, \tag{20}$$

$$\phi^0 = A(\xi, \tau)\left\{\frac{1}{1-c}\left[\frac{n+1}{2}\right]^{1/2}\psi_{n+1}(y) - \frac{1}{1+c}\left[\frac{n}{2}\right]^{1/2}\psi_{n-1}(y)\right\}, \tag{21}$$

where the $\psi_n(y)$ are the usual normalized Hermite functions, $\psi_n(y) = \exp(-[1/2]y^2)H_n(y)$, where $H_n(y)$ is the normalized Hermite polynomial of degree n.

Matching powers of ϵ gives the first order set:

$$-c_0 u_\xi^1 + yv^1 + \phi_\xi^1 = -u_\tau^0 - u^0 u_\xi^0 - v^0 u_y^0, \tag{22}$$

$$yv^1 + \phi_y^1 = c_0 v_\xi^1, \tag{23}$$

$$-c_0\phi_\tau^1 + u_\xi^1 + v_y^1 = -\phi_\tau^0 - (u^0\phi^0)_\xi - (v^0\phi^0)_y. \tag{24}$$

As in all perturbation theories, the left-hand side of the first order set is identical with that of the zeroth-order set. We can reduce this system down to the single equation

$$v_{yy}^1 + (-1/c_0 - y^2)v^1 = F_4, \tag{25}$$

where, identifying the right-hand sides of (5.14) as F_1, F_2 and F_3, respectively,

$$F_4 = [1 - 1/c_0^2](yF_1 + c_0 F_{2\xi}) + yF_1/c_0^2 + yF_3/c_0 + F_{1y}/c_0 + F_{2y}. \tag{26}$$

Because (25) has a homogeneous solution, the forced boundary value problem has a finite solution if and only if F_4 is orthogonal to the homogeneous solution, i.e.

$$\int_{-\infty}^{\infty} dy\psi_n(y) F_4(\xi, y, \tau) = 0. \tag{27}$$

The integration removes the y-dependence, but F_4 depends on ξ and τ through $A(\xi, \tau)$ and its *derivatives*. Since (27) must hold for *all* ξ and τ, this condition demands that $A(\xi, \tau)$ must satisfy a *differential* equation, which turns out to be the Korteweg–de Vries equation:

$$A_\tau + a(n)AA_\xi + b(n)A_{\xi\xi\xi} = 0. \tag{28}$$

Converting variables back to the earth-fixed coordinate system yields

$$A_t + c_0(n)A_x + a(n)AA_x + b(n)A_{xxx} = 0. \tag{29}$$

One striking feature of the KdV coefficients and also of the latitudinal structure functions given by (5.12) is that they are functions *only* of the *latitudinal mode number n*. Consequently, $a(n)$ and $b(n)$ for all possible cases can be summarized for the first 20 modes by table 1 of Boyd (1980b).

The explicit lowest order perturbative approximation to the lowest symmetric ($n = 1$) equatorial Rossby soliton is

$$u = A(x,t) \left(-9 + 6y^2\right) \exp\left(-(1/2)y^2\right),$$

$$\phi = A(x,t) \left(3 + 6y^2\right) \exp\left(-(1/2)y^2\right),$$

$$v = \frac{dA(x,t)}{dx}(2y) \exp\left(-(1/2)y^2\right), \tag{30}$$

where ϕ is the difference between the total depth h and its mean, $h_{mean} = 1$,

$$A(x,t) = 0.771\, B^2 \mathrm{sech}^2\{B[x - (-1/3 - 0.395B^2)t]\}. \tag{31}$$

The pseudowavenumber B thus controls both the longitudinal width and the amplitude of the solitary wave. Figure 4 shows the fields for a typical case. The $n = 1$ Rossby soliton is a pair of contra-rotating vortices, anticyclonic in each hemisphere, traveling at a faster than linear speed westward.

There is one modest complication. If we neglect the equatorial currents, we find that $a(n) = 0$ for all even n, i.e. for all modes that have u and ϕ antisymmetric about the equator. This does not imply that nonlinear effects are nonexistent

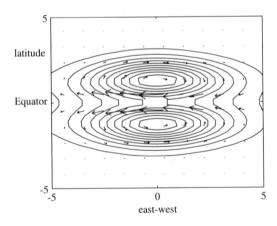

Figure 4: A typical equatorial Rossby solitary wave in the small-amplitude/KdV regime. The solid contours are of $\phi = h - 1$, which is also the eddy pressure. This solution was generated numerically, but is well approximated by the perturbative approximation for $B = 0.4$, which was used as the initial condition.

for these modes; they are merely weaker. Boyd shows that by extending the perturbation theory to one higher order, one can derive the so-called 'Modified' Korteweg–deVries (MKdV) equation (Boyd 1980b). This is identical in form to (29) except that the nonlinear term is $A^2 A_x$.

For several reasons, we will omit further discussion of the antisymmetric modes. First, nonlinear effects are weaker for these waves than for the symmetric modes. Second, explicit calculation shows that these equatorial waves do not form solitons: the MKdV coefficients are such that for equatorial Rossby waves, nonlinearity *accelerates* dispersion rather than opposing it. Third, the symmetric component of the equatorial forcing is considerably stronger than the antisymmetric component (except in the Indian Ocean). Fourth, the currents will invalidate the MKdV analysis by creating a small, nonzero coefficient $a(n)$ for the KdV equation. Unfortunately, this shear-induced current is sensitive to the mean flow profile.

One subtlety is hidden in the perturbative analysis: although the lowest ($n = 1$) latitudinal mode solitary waves are classical, decay-to-zero-at-spatial-infinity solitons, all the higher latitudinal mode solitary waves are weakly nonlocal, radiating small quasi-sinusoidal waves of lower latitudinal mode number. Because the strength of the radiation is proportional to $\exp(-q/B)$, where $q > 0$ is a constant and B is the pseudowavenumber (and also the square root of the amplitude of the core of the soliton), the radiation lies 'beyond all orders' in a perturbation series in powers of the small parameter B. Special methods of 'hyperasymptotic' perturbation theory are needed to understand this radiation (Boyd 1989b, 1991b; Segur *et al.* 1991; Boyd 1998a, 1998b).

This exponential-of-a-reciprocal dependence on the core amplitude is very common, but far from universal. Flierl and Petviashvili independently derived an equation for radially symmetric Rossby vortices in the middle latitudes which can be solved by a single, universal function of radius, the 'Flierl–Petviashvili monopole' (Flierl 1979; Petviashivili 1981; Boyd 1989a). Swenson showed that the monopole radiates at higher order in perturbation theory (Swenson 1986) and is therefore weakly nonlocal. For equatorial waves, however, the radiation is invisible to the standard multiples scales series even if calculated to the millionth power of B!

The effects of shear on nonlinear equatorial waves have not been thoroughly studied. Boyd (1984) and Greatbatch (1985) both treat shear using perturbation theory, but other studies such as Redekopp (1977) have demonstrated that the nonlinear/dispersive perturbation theory goes through with only minor changes even when the shear is incorporated directly into the zeroth-order equations and these are solved numerically.

The *qualitative* behavior of Rossby waves, however, will not be altered by the mean current because this merely alters the numerical coefficients $a(n)$ and $b(n)$. We note that at least in the absence of shear, the signs of the coefficients are such that equatorial Rossby and Kelvin solitary waves are *always positive*. In other words, these solitons are crests in the language of the shallow-water wave equations and thicken the upper layer by lowering the thermocline in the context of the 1–1/2 layer model. The phenomenology of the KdV equation is discussed elsewhere in this book and in many other sources such as (Boyd 1980b; Remoissenet 1991). Williams and

Wilson (1988) showed numerically that collisions of equatorial solitons of different heights are indeed highly elastic as predicted by KdV theory.

5.3 Equatorial modons

There are three ways to proceed beyond the lowest order, KdV-yielding perturbation theory. Boyd (1985) carried the expansions to the next order and showed that the first order corrections are still small compared to the zeroth-order terms even when the pseudowavenumber B (in a solution proportional at lowest order to sech$^2(B[x - ct])$) is sufficiently large ($B > 0.53$) so that recirculation occurs. That is, the wave does not merely propagate through the fluid, but instead a quasi-cylindrical disc of water is trapped within the travelling vortex pair.

The second strategy is to assume that the solution is a steadily propagating wave, traveling at a constant speed c, and then *numerically* solve the same nonlinear eigenvalue problem as solved *perturbatively* in the previous section:

$$(u - c)u_s + vu_y - yv + \phi_s = 0, \tag{32}$$

$$(u - c)v_s + vv_y + yu + \phi_y = 0, \tag{33}$$

$$-c\phi_s + u_s + v_y + (u\phi)_s + (v\phi)_y = 0, \tag{34}$$

where $s = x - ct$ is the east–west coordinate in a frame of reference moving with the wave (Boyd 2002a, 2002b). The third strategy is to solve the *inital value* problem for various initial conditions and see what develops. Kindle (1983) and Greatbatch (1985) generated solitons through excitation by wind stresses in oceans bounded by eastern and western coasts; Williams and Wilson (1988) and Boyd (2002b) used unforced initial value simulations in an unbounded sea to explore soliton–soliton collisions and large-amplitude Rossby solitary waves, respectively. Bouchut *et al.* showed that when a Rossby soliton propagates westward into shallower water [the opposite of the usual ocean situation, as the authors note, but still very illuminating], the Rossby wave grows in amplitude (since the same energy is moving a smaller volume of water) and transforms into a modon, i.e. the topography induces a region of recirculating fluid to be trapped within taller and narrower solitary waves (Bouchut *et al.* 2005).

Collectively, these studies show that the lowest latitudinal mode equatorial Rossby solitons are both KdV solitary wave and modon. This is very surprising because the mathematical ideas behind KdV theory and modon theory are radically different.

As illustrated above, the KdV model is derived for equatorial waves, as it has been for many other wave systems including some completely divorced from fluid mechanics, by a reductive perturbation theory that is simultaneously both a *long-wave* and a *small-amplitude* expansion. In contrast, there is no small parameter in modon theory. Indeed, an essential feature of modons is that they contain a disc of recirculating fluid, i.e. water that is trapped within the traveling vortex pair. Rather than being a perturbation of small amplitude, a modon can be a classical, nonradiating solitary wave only if its amplitude is *sufficiently large* – large enough so that the nonlinear interaction of the two vortices can accelerate the vortex pair beyond the range of linear phase speeds.

The analytical modons of Chaplygin and Lamb, generalized to include the beta-effect by Stern (1975) and Larichev and Reznik (1976), are exact finite amplitude solutions (Flierl et al. 1980), expressed in a polar coordinate system centered on the midpoint of the line connecting the two vortex centers. The region of recirculating fluid is bounded by a circle which is the outermost closed streakline, i.e. the largest streamline which does not extend to infinity in a coordinate system moving with the wave. The beta-plane solution is the product of $\sin(\theta)$, where θ is the angle in the polar coordinate system with Bessel functions of radius $- J_1$ in the region of closed streaklines, K_1 in the exterior. The potential vorticity is a linear function of the streakfunction everywhere, but with different slopes inside and outside the outermost closed streakline.

Some features of the special, *analytic* solutions are not general. Boyd and Ma (1990) and Eydeland and Turkington (1988) numerically computed modons in which the recirculating fluid is bounded by an ellipse or by two circular discs, respectively. These eigenvalue solutions and also some initial value experiments have yielded modons with nonlinear relationships between potential vorticity and stream function.

When the KdV approximations to the lowest equatorial Rossby soliton are extended to sufficiently large B that recirculation occurs, Boyd (1985) has shown that the approximation is still rather accurate, at least at the lower end of the recirculating regime, $B > 0.53$. However, these solitons have many characteristics in common with modons, including the following:

1. dipolar vorticity: two equal and opposite centers of rotation,
2. streaklines are symmetric about the north–south axis,
3. streaklines are antisymmetric with respect to the east–west axis,
4. the solitary wave propagates east–west only and remains centered on the same latitude,
5. the fluid inside the closed streaklines is bound to the wave, and must forever recirculate within the region circumscribed by the outermost closed streakline,
6. the phase speed is outside the range of speeds for a *linear* Rossby wave of any zonal scale,
7. the outermost closed streakline is a circle (or nearly circular),
8. the potential vorticity/streakfunction relation is linear in the exterior and either linear or nonlinear in the recirculation region.

The chief difference between midlatitude modons and equatorial $n = 1$ Rossby solitons is that the equatorial solitary wave is in the lowest latitudinal mode: it cannnot radiate Rossby waves of lower mode number when the soliton is weak because there are no lower modes to radiate to! Thus, the equatorial solitary wave is a classical soliton for a full range of amplitudes from very small to very large. In the large-amplitude, regime where recirculation occurs, it may be justly dubbed an 'equatorial modon'. And yet the KdV theory is a good approximation to this vortex pair, too.

5.4 Observations of equatorial Rossby solitons

Though solitons of KdV type have been readily generated and observed in numerical models, there have been no 'smoking guns' in the real ocean. Indeed, the best observations of significantly nonlinear equatorial waves are of Jupiter. So-called 'hot spots' with an east–west scale of 5,000–10,000 km, interspersed with 'plumes' of high cloud opacity, perhaps 10,000 km long, are observed at roughly six to eight degrees of latitude in the northern hemisphere of Jupiter. By comparing numerical simulations with the observations, Showman and Dowling (2000) concluded that the spots and plumes are nonlinear equatorial Rossby waves. When nonlinearity was omitted, the spots and plumes rapidly dispersed, but were long-lived when nonlinearity was restored. These observed waves do not have recirculation regions, and Showman and Dowling therefore identify them as equatorial cnoidal waves rather than solitary waves. As discussed above, however, a cnoidal wave *is* a train of weakly interacting solitary waves except at very small amplitude (where the overlap between neighboring solitons is large).

Boyd (2002b) speculates that in fact strong, equatorial Rossby solitary waves *have* been observed in the tropical ocean, but not identified as such. The large anticyclonic, slightly off-the-equator vortices that were first seen in satellite imagery 30 years ago are embedded in a shear zone between two strong mean jets, flowing in opposite directions. To analyze these so-called 'Legeckis eddies' or 'tropical instability waves', we first need to discuss the larger problem of a vortex or vortex train in a shear flow as will be done in Section 7. Before we turn to this, we will complete our discussion of planetary solitons near the equator by describing Kelvin solitary waves in the next section.

6 Equatorial Kelvin solitary waves

The Kelvin wave, which always propagates eastward at a phase speed triple that of the fastest linear Rossby wave, is the gravest mode of the equatorial ocean. Because it is symmetric with respect to the equator (as is much of the forcing for tropical waves) and because it is the lowest mode, it is excited more strongly than any other mode of the tropical ocean. All sorts of wind stresses, and reflections of waves from the western coast of an ocean, excite Kelvin waves on a variety of spatial and temporal scales. During an El Niño event, a weakening of the trade winds in the western Pacific triggers a massive upper-layer-thickening Kelvin wave that causes enormous warming of the eastern equatorial waters a couple of months later. The Kelvin wave then partially reflects as Rossby waves (which will form Rossby solitons on the leading edge) and partially as coastally trapped Kelvin waves that radiate north and south. The warm water triggers torrential coastal rains, and the houses of Hollywood movie stars, perched precariously on hills above Pacific beaches, are undercut and slide into the sea. More subtly, these same atmospheric changes modify the weather over the entire planet, creating the atmospheric phenomenon called the Southern Oscillation – a bad name, since both hemispheres are affected. The Southern Oscillation in turn modifies the trade winds, completing the link in a coupled ocean–atmosphere cycle with an irregular period of several years, which geophysicists now call ENSO ('El Niño/Southern Oscillation'). The Kelvin wave *is* El Niño; it is not an exaggeration to say that when the Kelvin wave is excited, the whole ocean/atmosphere system trembles. So it is obviously important to understand how nonlinearity reshapes this wave.

When there is no shear, Boyd (1980a) showed that the longitude and time factor of the Kelvin wave evolves according to the one-dimensional advection (ODA) equation, also known as the inviscid form of Burgers' equation:

$$A_t + A_x + 1.225AA_x = 0 \quad \text{[ODA equation]}. \tag{35}$$

(Later, independent derivations by different singular perturbation methods were given by Ripa (1982), Majda *et al.* (1999) and Le Sommer *et al.* (2004)). Because infinitesimal Kelvin waves all travel at unit nondimensional phase speed, there is no dispersion to balance nonlinear steepening, and the Kelvin wave in this model *always breaks* after a finite time. That is, the leading edge of the wave steepens to an infinite slope (The hydrostatic approximation inherent in the shallow-water equations breaks down at a very steep slope, and merely adding *longitudinal* viscosity, to make the ODA equation into Burgers' equation, is physically unsatisfactory since a very steep wave is likely to diffuse primarily in the *vertical*. The post-breaking behavior of equatorial Kelvin waves is therefore still a mystery.).

However, the equatorial ocean has a multiplicity of rather strong mean jets such as the South Equatorial Current (SEC), North Equatorial Countercurrent (NECC) and North Equatorial Current (NEC). Although these currents have only a quantitative effect on Rossby solitons (in the sense that solitary waves would still form even in the absence of these currents), the mean jets qualitatively alter the Kelvin wave by creating dispersion. This in turn can balance the nonlinear frontogenesis to create solitary waves, again described in longitude and time by the Korteweg–de Vries equation. As explained in (Boyd 1984; Long and Chang 1990), the shear-dependent coefficient of the third derivative in the KdV equation, appended to (35), can be calculated numerically by solving a *linear* eigenvalue problem with coefficients that depend on the mean east–west flow $U(y)$ (Boyd 2005b).

A typical Kelvin solitary wave in a numerical model is illustrated in Fig. 5. The zonal mean velocity and depth were

$$U(y) = -(1/2)\exp\left(-(25/64)y^2\right)\left(1 - \frac{25}{64}y^2\right)\left(1 + \frac{25}{36}y^2\right), \tag{36}$$

$$\Phi(y) = -(1/2)\exp\left(-(25/64)y^2\right)\left(\frac{512}{225} + \frac{25}{18}y^2 + \frac{25}{72}y^4\right). \tag{37}$$

To this was added an initial condition with $u = \phi$, as for a linear Kelvin wave in the absence of shear, and proportional to $\exp\left(-[1/2]y^2\right)$, as for a linear, shearless Kelvin wave, and with the x-dependent factor $0.12\text{sech}^2(0.7x)$. This was chosen to mimic the shape of a KdV solitary wave; the width and amplitude were adjusted empirically until the ensuing Kelvin pulse, after a bit of transient adjustment, remained unchanged over a very long period of time. In the absence of

Kelvin soliton

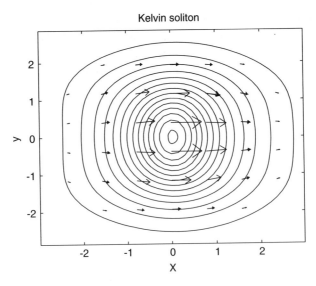

Figure 5: Contours of the wave height, $\phi \equiv h - H(y)$. The arrows show the direction and magnitude of the eddy current. The soliton is shown at $t = 100$ after integration from the initial condition $u(x, y, 0) = \phi(x, y, 0) = 0.12\mathrm{sech}^2(0.7x)\exp(-(1/2)y^2)$ with the mean currents specified in the main text. The computation grid was $x \in [-50, 50]$ with $y \in [-2\pi, 2\pi]$ using the eighth order model of (Boyd 1998a, 2002b).

shear, one can calculate using the formula of Boyd (1980a) that this initial condition will break at $t = 12.67$. The fact that the soliton is still roughly symmetric in the east–west about its crest at $t = 100$ shows that the shear-induced dispersion is indeed balancing nonlinearity.

The illustrated soliton was chosen to have almost identical east–west and north–south scales to demonstrate a theoretical challenge: although the nondimensional amplitude is small (the maximum wave height field is only about 12 % of the undisturbed depth), this particular solitary wave is not zonally elongated in the least. The KdV theory, which makes a long-wave approximation, *qualitatively* describes this solitary wave. However, one would really like something better, a theory that does not make the long-wave approximation, a sort of 'KdV-plus' theory.

The KdV theory predicts that Kelvin solitons of arbitrarily large height (and arbitrarily narrow width) are possible. However, this is false. The third derivative of the KdV equation implies that for an infinitesimal wave proportional to $\exp(ikx)$, this term – and therefore dispersion – will become infinite in the limit $k \to \infty$. In reality, dispersion actually goes to *zero* as $k \to \infty$, as shown analytically in (Boyd 2005b). It seems likely, on the basis of theory and numerical experiments, that the Kelvin wave exhibits the Cnoidal/Corner Wave/Breaking Scenario: solitary and cnoidal waves for small amplitude, breaking for large amplitude, and a travelling wave of finite maximum amplitude, which is a 'corner wave' in the sense that the longitudinal profile has a discontinuous first derivative at the crest of the wave (Boyd 2005a, 2006).

Because the shear-induced dispersion is weak, the amplitude of the corner wave is rather small. The importance of Kelvin solitary waves in the ocean is uncertain and no clear observations of such are known. However, it is a truism that the great progress in ocean observing systems has only moved the state of ocean data from the Stone Age to the Bronze Age. Perhaps, Kelvin solitary waves or cnoidal waves will be seen in the not too distant future.

7 Vortices embedded in a shear zone and nonlinear critical latitude theory

Westward propagating perturbations of the Equatorial Current System have been referred to as 'equatorial long waves' (Rossby gravity waves causing equatorial meanders of the EUC/SEC), 'Legeckis waves' (cusps in the north equatorial front, seen in satellite images). Our results, integrating from a rich range of sensors, indicate that away from the Equator, they consist of finite amplitude vortices, drifting along the shear layer between the SEC and the NECC. Although when sampled from a fixed point, they appear as periodic oscillations of the fields, just as linear waves would, they are clearly trains full of fully developed large-amplitude nonlinear eddies.

Flament et al. (1996).

Trains of anticyclonic eddies, embedded in the shear zone between the eastward-flowing North Equatorial Counter Current (NECC) and the westward-flowing South Equatorial Current (SEC), were discovered by Legeckis in satellite images nearly 30 years ago. The article by Flament *et al.* was the first study that used extensive *in situ* buoys to characterize these eddies. It is now well understood that these vortices form first as unstable waves that grow due to barotropic or mixed barotropic–baroclinic instability. At large amplitude, these waves roll up into nearly circular eddies. The vortices, after they have axisymmetrized, seem to evolve little further, propagating steadily like solitary waves (Weidman *et al.* 1999; Kennan and Flament 2000). (The most common name today seems to be 'tropical instability waves' or TIWs, even after the roll-up into vortices.)

These vortices bear many similarities to the $n = 1$ equatorial Rossby solitons described in an earlier section, but there are important differences too. The TIWs are anticyclones and do move westward relative to the average mean flow at their location. However, the mean currents are not symmmetric with respect to the equator and seem to suppress anticyclones in the southern hemisphere where the shear is not quite strong enough for instability. Without shear, the $n = 1$ solitons have vortex centers about three degrees off the equator; with shear, the TIW centers are forced a little further northward to about 4.4 N, right in the center of the shear zone.

Classical soliton theories, such as the usual KdV or NLS models, assume that the soliton forms by rearrangement of an existing wave disturbance rather than by the organization or reorganization of an unstable, amplifying wave at the expense of an initial mean flow. However, this exclusion of origin-by-instability is arguably rather narrow-minded. Figure 6 shows that the mature TIWs certainly look like propagating vortices. But the quasi-steady-state vortices have not been carefully studied for nonlinear/dispersive balances and other spoors of a soliton. Boyd's assertion that the mature vortices are best conceptualized as a cnoidal wave, a train of dynamically isolated solitary waves, remains only a plausible conjecture (Boyd 2002b).

The TIWs, alias Legeckis eddies, are but one example of the rather widespread phenomenon of vortices in a shear flow.

The most dramatic observed example of a vortex-in-shear is the Great Red Spot (GRS) of Jupiter, an anticyclonic vortex of elliptical shape, which has been spinning at about 20° S. for more than three centuries. Unlike the Legeckis eddies, the GRS is a single vortex: one of the major theoretical challenges has been to explain its singleness.

Ingersoll (1973) published the first suggestion that the GRS was a 'free vortex,' and therefore an unforced coherent structure. Maxworthy, Redekopp and Weidman (Maxworthy and Redekopp 1976a, 1976b; Redekopp 1977; Maxworthy *et al.* 1978; Redekopp and Weidman 1978) developed a KdV theory of GRS-as-Rossby soliton; the alternating jets north and south of the vortex create a waveguide (Dickinson 1968), just as Coriolis refraction creates the equatorial waveguide discussed earlier, and the Rossby dynamics is effectively one-dimensional.

The numerical simulations of Marcus (1990, 1993) and the laboratory experiments of Sommeria *et al.* (1991) showed that the GRS could form from a shear instability, just like the TIWs; in an atmosphere lacking the coastal boundaries of an ocean, the slow process of merger of like-signed vortices could eventually coalesce an initial train of vortices into a single mighty survivor.

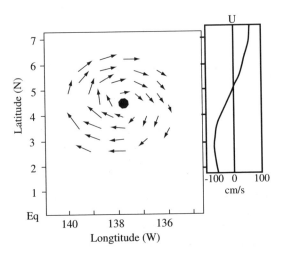

Figure 6: Observations of TIWs in their maturity. Left: arrow plot of the current vectors of the vortex. Right: the zonal mean east–west current in which the vortex is embedded. (Based on data from Flament *et al.* (1996).)

It is impossible to give a detailed discussion of all the theory of the vortex-in-shear because it is vast: all of nonlinear critical latitude theory is highly relevant. Superficially, this seems rather puzzling. When fluid waves are linearized about a mean flow $U(y)$, flowing in the same direction that the wave is propagating, the linearized wave equation will be *singular* at the 'critical latitude' y_c where $U(y_c) = c$. The singularity can be resolved without nonlinearity by either a viscous boundary layer around the critical layer (Dickinson 1968) or by transience, i.e. fluctuations in the strength of the forcing of the wave far from the critical latitude (Dickinson 1970). However, because the wave amplitude must rise steeply as the (linear, inviscid, steady) singularity is approached, it is very easy for nonlinearity to dominate in a thin layer around the critical latitude even if nonlinear effects are negligible everywhere else. This has inspired a long stream of papers on the 'nonlinear critical layer' beginning with the classic singular perturbation analyses of Benney and Bergeron (1969) and Davis (1969).

The reason that this is connected with the vortex-in-shear problem is that when the nonlinearity is important only in a thin layer, the layer consists of a chain of weakly nonlinear, zonally elongated vortices. This makes possible a small-amplitude-and-long-wave approximation just as in KdV theory! Conversely, the GRS theories of Redekopp, Weideman and Maxworthy assumed that the strong zonal shear was $O(1)$ and thus incorporated it into the coefficients of the linear eigenvalue problem, which determines the latitudinal structure of the KdV soliton. This linear equation is singular at the critical latitude; Redekopp (1977) and Redekopp and Weideman (1978) include a nonlinear critical layer analysis.

The early papers assumed a steady-state wave and mean flow, but later papers explored how the nonlinear critical layer would evolve over time. This turns out to be rather complicated because as isovorticity lines are wrapped around the intensifying vortices in the critical layer, the steepening gradients trigger local shear instabilities, and the core becomes very difficult to entangle even with supercomputers (Béland 1976; Warn and Warn 1978; Haynes 1989).

Stuart opened a nonperturbative line of attack by showing that the two-dimensional incompressible flow equations (without beta-effect) have an exact solution in the form of a periodic chain of same-signed vortices embedded in a shear flow,

$$\psi = -\log\left(C\cosh(\alpha y) + \sqrt{C^2 - 1}\cos(\alpha x)\right), \qquad (38)$$

where α and C are arbitrary constants with $C \geq 1$. Mallier (1995) extended the Stuart solutions to include beta using perturbation theory.

Figure 7 illustrates a typical Stuart vortex. The qualitative resemblance to Tropical Instability Waves and to the Great Red Spot is obvious.

Unfortunately, the existing theory is full of lacunae. Mallier, for example, treated both large and small β, but obtained explicit solutions only by using perturbation series in the nondimensional β for weak beta-effect. He suspects that even this series develops singularities at higher order.

The earlier and highly influential matched asymptotic expansions of Benney and Bergeron (1969) and Davis (1969) are reminiscent of the Sidney Harris cartoon

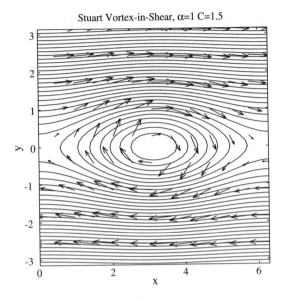

Figure 7: A vortex embedded in a shear flow. The contours are the streamfunction. The arrows show the direction and intensity of the flow. The graph is an instance of the Stuart vortex, but is typical of any vortex (or periodic chain of vortices) embedded in a shear zone.

where, in the middle of a long blackboard derivation, algebraic symbols are abruptly replaced by the chalked words 'A Miracle Happens!'. This is true in the sense that these authors found that systematic matching did not yield a unique solution. Benney and Bergeron therefore postulated that viscosity would homogenize the potential vorticity inside the vortex. However, this is only one possibility as explained by Mallier (1995). The time-dependent studies show that at least for small and moderate times, the interior of a critical layer has a complicated internal structure that depends on its history: spiral filaments of strong and weak vorticity wrapped around the center like layers of the Greek pastry, balaclava. Does the infinite-time homogenization postulated by Benney and Bergeron have any real connection with vortices in the real world, which are constantly forming and reforming as the wave forcing changes with passing storms and alternating seasons? Marcus (1993) argues on experimental and theoretical grounds for a nonhomogeneous vortex core for the GRS, despite its great age, with vorticity concentrated mostly around the periphery. What really happens on Jupiter and in the ocean? Does a solitary/cnoidal/KdV perspective on critical layer vortices enlighten or merely confuse? Can we trust a small-amplitude, long-wave model for the GRS when its zonal length is less than twice its latitudinal width, and its amplitude is rather large? And how, oh, how to describe the intense turbulence that flows around the edges of the GRS, bringing small vortices for the GRS to eat by merger, squeezed against the GRS by the mean jets north and south of it?

Derzho and Grimshaw, in a study of *large*-amplitude Rossby solitons in a channel, show that clever expansions may yet yield new insights into solitary waves (2005). Even so, familiarity with perturbation theory is evaporating from atmospheric, oceanic and mechanical engineering departments like window frost on a sunny day. Do the 'code cowboys' have the interest and means to sort out the vortex-in-shear problem when the 'perturbers' are all white-haired?

8 Summary

In this brief review, we have necessarily omitted many topics. The numerical computation of traveling Rossby solitons by solving a nonlinear eigenvalue problem, for instance, is still very challenging. However, some very interesting special-purpose algorithms have been developed as described in (Petviashvili 1976; Eydeland and Turkington 1988; Vallis *et al.* 1989, 1990; Carnevale and Vallis 1990; Shepherd 1990; Turkington *et al.* 1991; Verkeley 1993; Haupt *et al.* 1993; Turkington and Whitaker 1996; Meiron *et al.* 2000; Rossi *et al.* 2002).

In spite of the vast body of work reviewed above, there are many unresolved issues. Some have already been mentioned above. For example, Boyd (2002b) has shown that there are at least two coexisting branches of lowest latitudinal mode equatorial solitons – both modon-like, but one with a linear potential vorticity/streakfunction relationship in the interior and the other with a strongly nonlinear relationship. Numerical solutions, both initial value and eigenvalue, fail to yield equatorial solitons above a rather large amplitude threshold, but how the branch ends, or turns at a limit point, and how many distinct branches of equatorial modons exist, is not presently understood.

Similarly, Rossby solitons embedded in a shear flow is still full of mysteries, even after four decades. How general can the vorticity distribution within the vortices be? What distributions are stable, and how does the vorticity rearrange itself when the vorticity is initially unstable? Are the embedded solitary waves truly steady state, or is there a slow, almost imperceptible evolution?

Lastly, good observations of Rossby solitary waves are still limited. There are certainly coherent structures of planetary scales, but for which is the primary balance between nonlinearity and dispersion? What periodic trains of vortices, such as TIWs, are best modeled as isolated vortices, and which are better understood as perturbed cosine waves, or does this distinction even matter?

Rossby and Kelvin solitons are an old field with 40 years of answers, but also a frontier field with just as many questions.

Acknowledgments

I thank the editor, Roger Grimshaw, for inviting me to write this review, and for pointing me to the work of Zeitlin and collaborators. This work was supported by the National Science Foundation through grant OCE0451951.

Bibliography

Barcilon, A. & Drazin, P.G., Nonlinear waves of vorticity. *Stud. Appl. Math.*, **106(4)**, pp. 437–479, 2001.

Béland, M., Numerical study of nonlinear Rossby wave critical layer development in a barotropic flow. *J. Atmos. Sci.*, **33**, pp. 2066–2078, 1976.

Bender, C.M. & Orszag, S.A., *Advanced Mathematical Methods for Scientists and Engineers*, McGraw-Hill: New York, 1978.

Benney, D.J. & Bergeron, R.F., A new class of nonlinear waves in parallel flows. *Stud. Appl. Math.*, **48**, pp. 181–204, 1969.

Bouchut, F., Le Sommer, J. & Zeitlin, V., Breaking of balanced and unbalanced equatorial waves. *Chaos*, **15(1)**, 013503, 2005.

Boyd, J.P., The nonlinear equatorial Kelvin wave. *J. Phys. Oceanogr.*, **10**, pp. 1–11, 1980a.

Boyd, J.P., Equatorial solitary waves. Part I: Rossby solitons. *J. Phys. Oceanogr.*, **10**, pp. 1699–1718, 1980b.

Boyd, J.P., Equatorial solitary waves. Part II: Envelope solitons. *J. Phys. Oceanogr.*, **13**, pp. 428–449, 1983.

Boyd, J.P., Equatorial solitary waves. Part IV: Kelvin solitons in a shear flow. *Dyn. Atmos. Oceans*, **8**, pp. 173–184, 1984.

Boyd, J.P., Equatorial solitary waves. Part III: Modons. *J. Phys. Oceanogr.*, **15**, pp. 46–54, 1985.

Boyd, J.P., New directions in solitons and nonlinear periodic waves: Polycnoidal waves, imbricated solitons, weakly non-local solitary waves and numerical boundary value algorithms. *Advances in Applied Mechanics*, Vol. 27, eds. T.Y. Wu & J.W. Hutchinson, Academic Press: New York, pp. 1–82, 1989a.

Boyd, J.P., Non-local equatorial solitary waves. *Mesoscale/Synoptic Coherent Structures in Geophysical Turbulence: Proc. 20th Liege Coll. on Hydrodynamics*, eds. J.C.J. Nihoul & B.M. Jamart, Elsevier: Amsterdam, pp. 103–112, 1989b.

Boyd, J.P., Nonlinear equatorial waves. *Nonlinear Topics of Ocean Physics: Fermi Summer School, Course LIX*, ed. A.R. Osborne, North-Holland: Amsterdam, pp. 51–97, 1991a.

Boyd, J.P., Weakly nonlocal solitary waves. *Nonlinear Topics of Ocean Physics: Fermi Summer School, Course LIX*, ed. A.R. Osborne, North-Holland: Amsterdam, pp. 527–556, 1991b.

Boyd, J.P., Nonlocal modons on the beta-plane. *Geophys. Astrophys. Fluid Dyn.*, **75**, pp. 163–182, 1994.

Boyd, J.P., High order models for the nonlinear shallow water wave equations on the equatorial beta-plane with application to Kelvin wave frontogenesis. *Dyn. Atmos. Oceans*, **28(2)**, pp. 69–91, 1998a.

Boyd, J.P., *Weakly Nonlocal Solitary Waves and Beyond-All-Orders Asymptotics: Generalized Solitons and Hyperasymptotic Perturbation Theory*, Mathematics and Its Applications, Vol. 442, Kluwer: Amsterdam, 1998b.

Boyd, J.P., The devil's invention: Asymptotics, superasymptotics and hyperasymptotics. *Acta Applicandae*, **56(1)**, pp. 1–98, 1999.

Boyd, J.P., Deleted residuals, the QR-factored Newton iteration, and other methods for formally overdetermined determinate discretizations of nonlinear eigen-problems for solitary, cnoidal, and shock waves. *J. Comput. Phys.*, **179(1)**, pp. 216–237, 2002a.

Boyd, J.P., Equatorial solitary waves. Part V: Initial value experiments, coexisting branches and tilted-pair instability. *J. Phys. Oceangr.*, **32(9)**, pp. 2589–2602, 2002b.

Boyd, J.P., Hyperasymptotics and the linear boundary layer problem: Why asymptotic series diverge. *SIAM Rev.*, **47(3)**, pp. 553–575, 2005.

Boyd, J.P., The cnoidal wave/corner wave/breaking wave scenario: a one-sided infinite-dimension bifurcation. *Math. Comput. Simulation*, **69(3–4)**, pp. 235–242, 2005a.

Boyd, J.P., The short-wave limit of linear equatorial Kelvin waves in a shear flow. *J. Phys. Oceangr.*, **35(6)**, pp. 1138–1142, 2005b.

Boyd, J.P., Fourier pseudospectral method with Kepler mapping for travelling waves with discontinuous slope: Application to corner waves of the Ostrovsky-Hunter equation and equatorial Kelvin waves in the four-mode approximation. *Applied Mathematics and Computation*, **177(1)**, pp. 289–299, 2006.

Boyd, J.P. & Haupt, S.E., Polycnoidal waves: spatially periodic generalizations of multiple solitary waves. *Nonlinear Topics of Ocean Physics: Fermi Summer School, Course LIX*, ed. A.R. Osborne, North-Holland: Amsterdam, pp. 827–856, 1991.

Boyd, J.P. & Ma, H., Numerical study of elliptical modons by a spectral method. *J. Fluid Mech.*, **221**, pp. 597–611, 1990.

Carnevale, G.F. & Vallis, G.K., Pseudo-advective relaxation to stable states of inviscid two-dimensional fluids. *J. Fluid Mech.*, **213**, pp. 549–571, 1990.

Carton, X., Hydrodynamical modeling of oceanic vortices. *Surveys Geophys.*, **22(3)**, pp. 179–263, 2001.

Clarke, R.A., Solitary and cnoidal planetary waves. *Geophysical Fluid Dynamics*, **2**, pp. 343–354, 1971.

Davis, R.E., On the high Reynolds number flow over a wavy boundary. *J. Fluid Mech.*, **36**, pp. 337–346, 1969.

de Szoeke, R.A., An effect of the thermobaric nonlinearity of the equation of state: a mechanism for sustaining solitary Rossby waves. *J. Phys. Oceanogr.*, **34(9)**, pp. 2042–2056, 2004.

Derzho, O.G. & Grimshaw, R., Rossby waves on a shear flow with recirculation cores. *Stud. Appl. Math.*, **115(4)**, pp. 387–403, 2005.

Dickinson, R.E., Planetary Rossby waves propagating vertically through weak westerly wind wave guides. *J. Atmos. Sci.*, **25(6)**, pp. 984–1002, 1968.

Dickinson, R.E., Development of a Rossby wave critical level. *J. Atmos. Sci.*, **27**, pp. 627–633, 1970.

Eydeland, A. & Turkington, B., A computational method of solving free-boundary problems in vortex dynamics. *J. Comput. Phys.*, **78**, pp. 194–214, 1988.

Fedorov, K.N. & Ginsburg, A.I., Mushroom-like currents (vortex dipoles) in the ocean and a laboratory tank. *Annales Geophys.*, **4**, pp. 507–516, 1986.

Flament, P.J., Kennan, S.C., Knox, R.A., Niiler, P.P. & Bernstein, R.L., The three-dimensional structure of an upper ocean vortex in the tropical Pacific Ocean. *Nature*, **382**, pp. 610–613, 1996.

Flierl, G.R., Baroclinic solitary waves with radial symmetry. *Dyn. Atmos. Oceans*, **3**, pp. 15–38, 1979.

Flierl, G.R., Isolated eddy models in geophysics. *Ann. Rev. Fluid Mech.*, **19**, pp. 493–530, 1987.

Flierl, G.R., Rings: semicoherent oceanic features. *Chaos*, **4**, pp. 355–367, 1994.

Flierl, G.R. & Haines, K., The decay of modons due to Rossby-wave radiation. *Phys. Fluids*, **6(10)**, pp. 3489–3497, 1994.

Flierl, G.R., Larichev, V.D., McWilliams, J.C. & Reznik, G.M., The dynamics of baroclinic and barotropic solitary eddies. *Dyn. Atmos. Oceans*, **5**, pp. 1–41, 1980.

Fornberg, B., A numerical study of 2-D turbulence. *J. Comput. Phys.*, **25**, pp. 1–31, 1977.

Gottwald, G. & Grimshaw, R., The formation of coherent structures in the context of blocking. *J. Atmos. Sci.*, **56(21)**, pp. 3640–3662, 1999.

Greatbatch, R.J., Kelvin wave fronts, Rossby solitary waves and nonlinear spinup of the equatorial oceans. *J. Geophys. Res.*, **90**, pp. 9097–9107, 1985.

Haines, K. & Marshall, J., Eddy-forced coherent structures as a prototype of atmospheric blocking. *Quart. J. Roy. Meteor. Soc.*, **113(476)**, pp. 681–704, 1987.

Haupt, S.E., McWilliams, J.C. & Tribbia, J.J., Modons in shear flow. *J. Atmos. Sci.*, **50(9)**, pp. 1181–1198, 1993.

Haynes, P.H., The effect of barotropic instability on the nonlinear evolution of a Rossby-wave critical layer. *J. Fluid Mech.*, **207**, pp. 231–266, 1989.

Ho, C.M. & Huerre, P., Perturbed free shear layers. *Ann. Rev. Fluid Mech.*, **16**, pp. 365–424, 1984.

Hopfinger, E.J. & van Heijst, G.J.F., Vortices in rotating fluids. *Ann. Rev. Fluid Mech.*, **25**, pp. 241–289, 1991.

Huang, F. & Lou, S.Y., Analytical investigation of Rossby waves in atmospheric dynamics. *Phys. Lett. A*, **320(5–6)**, pp. 428–437, 2004.

Ingersoll, A.P., Jupiter's Great Red Spot: A free atmospheric vortex? *Science*, **182**, pp. 1346–1348, 1973.

Ingersoll, A.P., Atmospheres of the giant planets (Part 15). *The New Solar System*, 4th edn, eds. J.K. Beatty, C.C. Petersen & A. Chaikin, Cambridge University Press: New York, pp. 201–220, 1999.

Ingersoll, A.P., Atmospheric dynamics of the outer planets. *Meteorology at the Millenium*, ed. R.P. Pearce, Academic Press: New York, pp. 306–315, 2002.

Jain, R.K., Goswami, B.N., Satyan, V. & Keshavamurty, R.N., Envelope soliton solution for finite-amplitude equatorial waves. *Proc. Indian Acad. Sci. – Earth and Planetary Sci.*, **90(3)**, pp. 305–326, 1981.

Kennan, S.C. & Flament, P.J., Observations of a tropical instability vortex. *J. Phys. Oceangr.*, **30(9)**, pp. 2277–2301, 2000.

Kindle, J., On the generation of Rossby solitons during El Nino. *Hydrodynamics of the Equatorial Ocean*, ed. J.C.J. Nihoul, Elsevier: Amsterdam, pp. 353–368, 1983.

Korotaev, G.K., Radiating vortices in geophysical fluid dynamics. *Surveys Geophys.*, **18(6)**, pp. 567–619, 1997.

Larichev, V.D. & Reznik, G.M., Two-dimensional solitary Rossby waves: an exact solution. *Dok Akad Nauk SSSR*, **231(5)**, pp. 1077–1079, 1976.

Le Sommer, J., Reznik, G.M. & Zeitlin, V., Nonlinear geostrophic adjustment of long-wave disturbances in the shallow-water model on the equatorial beta-plane. *J. Fluid Mech.*, **515**, pp. 135–170, 2004.

Lilly, D.K., Numerical simulation of two-dimensional turbulence. *Phys. Fluids*, **12(Suppl. II)**, pp. 240–249, 1969.

Lilly, D.K., Numerical simulation of developing and decaying two-dimensional turbulence. *J. Fluid Mech.*, **45(2)**, pp. 395–415, 1971.

Long, B. & Chang, P., Propagation of an equatorial Kelvin wave in a varying thermocline. *J. Phys. Oceanogr.*, **20**, pp. 1826–1841, 1990.

Luo, D.H., Planetary-scale baroclinic envelope Rossby solitons in a two-layer model and their interaction with synoptic-scale eddies. *Dyn. Atmos. Oceans*, **32(1)**, pp. 27–74, 2000.

Majda, A.J., Rosales, R.R., Tabak, E.G. & Turner, C.V., Interaction of large-scale equatorial waves and dispersion of Kelvin waves through topographic resonances. *J. Atmos. Sci.*, **56(24)**, pp. 4118–4133, 1999.

Malanotte Rizzoli, P., Planetary solitary waves in geophysical flows. *Advances in Geophysics*, **24**, pp. 147–224, 1982.

Malanotte Rizzoli, P. & Hancock, P.J., Nonlinear stationary Rossby waves on nonuniform zonal winds and atmospheric blocking. Part 4: A comparison between theory and data. *J. Atmos. Sci.*, **44(17)**, pp. 2506–2529, 1987.

Malguzzi, P. & Malanotte Rizzoli, P., Nonlinear stationary Rossby waves on nonuniform zonal winds and atmospheric blocking. 1. The analytical theory. *J. Atmos. Sci.*, **41(17)**, pp. 2620–2628, 1984.

Mallier, R., Stuart vortices on a beta-plane. *Dyn. Atmos. Oceans*, **22**, pp. 213–238, 1995.

Marcus, P.J., Vortex dynamics in a shearing zonal flow. *J. Fluid Mech.*, **215**, pp. 393–430, 1990.

Marcus, P.S., Jupiter's Great Red Spot and other vortices. *Ann. Rev. Astron. Astrophys.*, **31**, pp. 523–573, 1993.

Marshall, H.G. & Boyd, J.P., Solitons in a continuously stratified equatorial ocean. *J. Phys. Oceanogr.*, **17**, pp. 1016–1031, 1987.

Maslowe, S.A., Critical layers in shear flows. *Ann. Rev. Fluid Mech.*, **18**, pp. 405–432, 1986.

Maxworthy, T. & Redekopp, L.G., New theory of the Great Red Spot from solitary waves in the Jovian atmosphere. *Nature*, **260**, pp. 509–511, 1976a.

Maxworthy, T. & Redekopp, L.G., A solitary wave theory of the Great Red Spot and other observed features in the Jovian atmosphere. *Icarus*, **29**, pp. 261–271, 1976b.

Maxworthy, T., Redekopp, L.G. & Weidman, P.D., On the production and interaction of planetary solitary waves: applications to the Jovian atmosphere. *Icarus*, **33**, pp. 388–409, 1978.

McWilliams, J.C., An application of equivalent modons to atmospheric blocking. *Dyn. Atmos. Oceans*, **5(1)**, pp. 43–66, 1980.

McWilliams, J.C., Interactions of isolated vortices. Part 2: Modon generation by monopole collision. *Geophys. Astro Fluid Dyn.*, **24(1)**, pp. 1–22, 1983.

McWilliams, J.C., The emergence of isolated coherent vortices in turbulent flow. *J. Fluid Mech.*, **146**, pp. 21–43, 1984.

McWilliams, J.C., Submesoscale coherent vortices in the ocean. *Rev. Geophys.*, **23(2)**, pp. 165–182, 1985.

McWilliams, J.C., Geostrophic vortices. *Nonlinear Topics of Ocean Physics: Fermi Summer School, Course LIX*, ed. A.R. Osborne, North-Holland: Amsterdam, pp. 5–50, 1991.

McWilliams, J.C., Phenomenological hunts in two-dimensional and stably stratified turbulence. *Atmospheric Turbulence and Mesoscale Meteorology*, eds. E. Fedorovich, R. Rotunno & B. Stevens, Cambridge University Press: Cambridge, pp. 35–49, 2004.

Medvedev, S.B. & Zeitlin, V., Weak turbulence of short equatorial waves. *Phys. Lett. A*, **342(3)**, pp. 217–227, 2005.

Meiron, D.I., Moore, D.W. & Pullin, D.I., On steady compressible flows with compact vorticity; the compressible Stuart vortex. *J. Fluid Mech.*, **409**, pp. 29–49, 2000.

Meleshko, V.V. & van Heijst, G.J.F., On Chaplygin's investigations of two-dimensional vortex structures in an inviscid fluid. *J. Fluid Mech.*, **272**, pp. 157–182, 1992.

Miles, J.W., Solitary waves. *Ann. Reviews Fluid Mech.*, **12**, pp. 11–43, 1980.

Nayfeh, A.H., *Perturbation Methods*, Wiley: New York, 1973.

Neven, E.C., Modons in shear flow on a sphere. *Geophys. Astrophy. Fluid Dyn.*, **74(1–4)**, pp. 51–71, 1994.

Nezlin, M.V. & Snezhkin, E.N., *Rossby Vortices, Spiral Structures, Solitons*, Springer-Verlag: New York, 1993.

Nezlin, M.V. & Sutyrin, G.G., Problems of simulation of large, long-lived vortices in the atmospheres of the giant planets (Jupiter, Saturn, Neptune). *Surveys Geophys.*, **15(1)**, pp. 63–99, 1994.

Nihoul, J.C.J. & Jamart, B.M., (eds.), *International Liége Colloquium on Ocean Hydrodynamics*, Vol. 20, Liége Colloquium on Ocean Hydrodynamics, Elsevier: Amsterdam, 1989.

Olson, D.B., Rings in the ocean. *Ann. Rev. Earth Planet Sci.*, **38**, pp. 283–311, 1991.

Petviashvili, V.I., Equation of an extraordinary soliton. *Soviet Journal of Plasma Physics*, **2**, pp. 257–260, 1976.

Petviashvili, V.I., Red Spot of Jupiter and the drift soliton in a plasma. *Soviet Physics JETP Letters*, **32**, pp. 619–622, 1981.

Pokhotelov, O.A., Kaladze, T.D., Shukla, P.K. & Stenflo, L., Three-dimensional solitary vortex structures in the upper atmosphere. *Physica Scripta*, **64(3)**, pp. 245–252, 2001.

Redekopp, L.G., Theory of solitary Rossby waves. *J. Fluid Mech.*, **82**, pp. 725–745, 1977.

Redekopp, L.G. & Weidman, P.D., Solitary Rossby waves in zonal shear flows and their interactions. *J. Atmos. Sci.*, **35**, pp. 790–804, 1978.

Remoissenet, M., *Waves Called Solitons: Concepts and Experiments*, 3rd edn, Springer-Verlag: New York, 1991.

Reznik, G.M. & Zeitlin, V., Resonant excitation of Rossby waves in the equatorial waveguide and their nonlinear evolution. *Phys. Rev. Lett.*, **96(3)**, 034502, 2006.

Ripa, P., Nonlinear wave-wave interactions in a one-layer reduced-gravity model on the equatorial beta-plane. *J. Phys. Oceangr.*, **12**, pp. 97–111, 1982.

Rossi, L.F. & Graham-Eagle, J., On the existence of two-dimensional, localized rotating, self-similar vortical structures. *SIAM J. Appl. Math.*, **62(6)**, pp. 2114–2128, 2002.

Segur, H., Tanveer, S. & Levine, H., (eds.), *Asymptotics Beyond All Orders*, Plenum: New York, 1991.

Shepherd, T.G., A general method for finding extremal energy states of Hamiltonian dynamical systems, with applications to perfect fluid. *J. Fluid Mech.*, **213**, pp. 573–587, 1990.

Shivamoggi, B.K., A generalized class of nonlinear Rossby localized structures in geophysical fluid flows. *Chaos, Solitons and Fractals*, **14(3)**, pp. 469–477, 2002.

Showman, A.P. & Dowling, T.E., Nonlinear simulations of Jupiter's 5-μm hot spots. *Science*, **289(5485)**, pp. 1737–1740, 2000.

Sommeria, J., Meyers, S.D. & Swinney, H.L., Experiments on vortices and Rossby waves in eastward and westward jets. *Nonlinear Topics of Ocean Physics: Fermi Summer School, Course LIX*, ed. A.R. Osborne, North-Holland: Amsterdam, pp. 227–269, 1991.

Spineau, F., Vlad, M., Itoh, K., Sanuki, H. & Itoh, S.I., Minimal properties of planetary eddies. *Phys. Rev. Lett.*, **93(2)**, 025001, 2004.

Stern, M.E., Minimal properties of planetary eddies. *Journal of Marine Research*, **33**, pp. 1–13, 1975.

Swenson, M., A note on baroclinic solitary waves with radial symmetry. *Dyn. Atmos. Oceans*, **10(3)**, pp. 243–252, 1986.

Toda, M., Studies of a non-linear lattice. *Physics Reports*, **18**, pp. 1–124, 1975.

Tribbia, J.J., Modons in spherical geometry. *Geophys. Astrophys. Fluid Dyn.*, **30**, pp. 131–168, 1984.

Turkington, B. & Whitaker, N., Statistical equilibrium computations of coherent structures in turbulent shear layers. *SIAM J. Sci. Comput.*, **17(6)**, pp. 1414–1433, 1996.

Turkington, B., Eydeland, A. & Wang, S., A computational method for solitary internal waves in a continuously stratified fluid. *Stud. Appl. Math.*, **85**, pp. 93–127, 1991.

Vallis, G.K., Carnevale, G.F. & Young, W.R., Extremal energy properties and construction of stable solutions of the Euler equations. *J. Fluid Mech.*, **207**, pp. 133–152, 1989.

Vallis, G.K., Carnevale, G.F. & Shepherd, T.G., A natural method for finding stable states of Hamiltonian systems. *Proceedings of the IUTAM Conference on Topological Fluid Dynamics*, ed. H.K. Moffatt, Cambridge University Press: New York, pp. 429–439, 1990.

Vasavada, A.R. & Showman, A.P., Jovian atmospheric dynamics: an update after Galileo and Cassini. *Repts. Prog. Phys.*, **68(8)**, pp. 1935–1996, 2005.

Verkley, W.T.M., The construction of barotropic modons on a sphere. *J. Atmos. Sci.*, **41**, pp. 2492–2504, 1984.

Verkley, W.T.M., Stationary barotropic modons in westerly background flows. *J. Atmos. Sci.*, **44**, pp. 2383–2398, 1987.

Verkley, W.T.M., Modons with uniform absolute vorticity. *J. Atmos. Sci.*, **47**, pp. 727–745, 1990.

Verkley, W.T.M., A numerical method to find form-preserving free solutions of the barotropic vorticity equation on a sphere. *J. Atmos. Sci.*, **50**, pp. 1488–1503, 1993.

Warn, T. & Warn, H., The evolution of a nonlinear critical level. *Stud. Appl. Math.*, **59**, pp. 37–71, 1978.

Weidman, P.D., Mickler, D.L., Dayyani, B. & Born, G.H., Analysis of Legeckis eddies in the near-equatorial Pacific. *J. Geophys. Res.*, **104(C4)**, pp. 7865–7887, 1999.

Whittaker, E.T. & Watson, G.N., *A Course of Modern Analysis*, 4th edn., Cambridge University Press: Cambridge, 1940.

Williams, G.P., Jovian dynamics. Part 1: vortex stability, structure and genesis. *J. Atmos. Sci.*, **53(18)**, pp. 2685–2734, 1996.

Williams, G.P. & Wilson, R.J., The stability and genesis of Rossby vortices. *J. Atmos. Sci.*, **45**, pp. 207–249, 1988.

Wu, R., Long wave approximation, linear and non-linear Rossby waves. *Scientia Sinica*, **29(3)**, pp. 302–312, 1986.

Yano, J.I. & Tsujimura, Y.N., The domain of validity of the KdV-type solitary Rossby waves in the shallow water beta-plane model. *Dyn. Atmos. Oceans*, **11(2)**, pp. 101–129, 1987.

Zhao, Q., Fu, Z.T. & Liu, S.K., Equatorial envelope Rossby solitons in a shear flow. *Adv. Atmos. Sci.*, **18(3)**, pp. 418–428, 2001a.

Zhao, S.N., Xiong, X.Y., Hu, F. & Zhu, J., Rotating annulus experiment: Large-scale helical soliton in the atmosphere? *Phys. Rev. E.*, **64(5)**, 056621, 2001b.

CHAPTER 7

Envelope solitary waves

R.H.J. Grimshaw
*Department of Mathematical Sciences, Loughborough University,
Loughborough, UK.*

Abstract

In this chapter, we discuss envelope solitary waves that arise when a plane periodic wave is amplitude modulated. In general, the modulation leads to a rapidly
oscillating carrier wave with a slowly varying envelope. In the weakly nonlinear
regime, envelope waves are described by the nonlinear Schrödinger equation when
attention is confined to just one spatial dimension. We shall focus our attention
on capillary–gravity waves, although envelope waves can arise in many other fluid
flow situations. When the nonlinear Schrödinger equation is of the focusing type,
it supports solitary waves (solitons). However, these solitary waves do not lead to
a steady solution of the full physical system, as the carrier's phase speed and the
envelope's group velocity are not usually equal. But in the special circumstances
when the phase and group velocities do coincide, there is a bifurcation to a steady
envelope solitary wave. Our main interest is in capillary–gravity waves, and for
these we give many details. Extensions to waves in more than one space dimension
is based on the Benney–Roskes system, and leads to 'lump' solitary waves.

1 Linear waves and group velocity

As noted in Chapter 1, there are two classes of solitary waves of interest, each
of which can be regarded as a bifurcation from those points in the linear spectrum where the group velocity and phase velocity are equal. When this bifurcation
occurs in the long-wave limit, it leads typically to the Korteweg–de Vries equation
and its solitary wave solutions. These have been described in detail in the preceding
chapters. In this chapter, we will describe the alternative situation when the bifurcation occurs at a finite non-zero wavenumber. In this case, the outcome for weakly

nonlinear waves is related to the nonlinear Schrödinger (NLS) equation and leads to envelope solitary waves; these waves have an underlying rapidly varying oscillation, contained with a smoothly varying envelope; consequently, their decaying tails are also oscillatory.

However, before coming to the specific case of envelope solitary waves, it is instructive to consider wave envelopes in a more general setting. Let us consider a fluid flow supporting waves and move to the linearized $\mathbf{x} = (x, y)$. Then a sinusoidal wave can be described by

$$\eta(x, t) = a \exp(i\mathbf{k} \cdot \mathbf{x} - i\omega t) + \text{c.c.} \qquad (1)$$

Here, η is one of the relevant physical flow variables, $\mathbf{k}$ is the wavenumber vector, ω is the wave frequency, while a is the constant amplitude and ϕ is a constant phase. This sinusoidal wave has a wavelength $\lambda = 2\pi/\kappa$, where $\kappa = |\mathbf{k}|$ is the magnitude of the wavenumber vector (for instance, in two spatial dimensions, $\mathbf{k} = (k, l)$ and so $\kappa = \sqrt{k^2 + l^2}$) and the wave period $T = 2\pi/\omega$; a is the (complex) amplitude and c.c. denotes the complex conjugate. The phase velocity has a magnitude $c = \omega/\kappa$ and is in the direction of $\mathbf{k}$. Equation (1) is a *kinematic* expression, valid for all physical systems that support waves. The *dynamics* of the fluid flow system in question will lead to the dispersion relation,

$$\omega = \omega(\mathbf{k}), \qquad (2)$$

defining the frequency as a function of wavenumber. The phase velocity $c = \omega/\kappa$ is likewise a function of wavenumber. Here we consider only stable waves for which ω is real-valued for all real wavenumbers $\mathbf{k}$. For instance, for water waves, the dispersion relation is (see, for instance, Lamb 1932; Mei 1983),

$$\omega^2 = g\kappa(1 + Bq^2)\tanh q, \quad q = \kappa h, \qquad (3)$$

where the Bond number $B = \Sigma/gh^2$ ($\rho\Sigma$ is the coefficient of surface tension and ρ is the water density, which has a value of 74 dynes/cm at 20°C).

The concept of group velocity arises when one considers a linear superposition of waves of several wavenumbers and corresponding frequencies. For simplicity, we consider the one-dimensional case when $\mathbf{k} = (k, 0)$ and η depends only on x and t. Thus, in general, we can replace (1) with

$$\eta(x, t) = \int_{-\infty}^{\infty} F(k) \exp(ikx - i\omega(k)t) \, dk, \qquad (4)$$

where $F(k)$ is the Fourier transform of $\eta(x, 0)$. Thus, the initial conditions determine the Fourier transform, each component of which evolves independently with frequency ω related to the wavenumber k through the dispersion relation $\omega = \omega(k)$ (the one-dimensional version of (2)). To obtain a wave packet, we suppose that the initial conditions are such that $F(k)$ has a dominant component centred at $k = k_0$. The dispersion relation is then approximated by

$$\omega = \omega_0 + c_g(k - k_0) + \delta(k - k_0)^2, \qquad (5)$$

where

$$c_g = \frac{d\omega}{dk} \quad \text{and} \quad \delta = \frac{1}{2}\frac{d^2\omega}{dk^2}. \tag{6}$$

Here, both c_g and δ are evaluated at $k = k_0$. The expression (4) then becomes

$$\eta(x, t) \approx A(x, t) \exp\left(i(k_0 x - \omega_0 t)\right) + \text{c.c.}, \tag{7}$$

where

$$A(x, t) = \int_{-\infty}^{\infty} F(k_0 + \kappa) \exp\left(i(\kappa(x - c_g t)) - i\delta\kappa^2 t\right) d\kappa, \tag{8}$$

and the variable of integration has been changed from k to $\kappa = k - k_0$. Here, the sinusoidal factor $\exp\left(i(k_0 x - \omega_0 t)\right)$ is a carrier wave with a phase velocity ω_0/k_0, while the (complex) amplitude $A(x, t)$ describes the wave packet. Since the term proportional to κ^2 in the exponent in (8) is a small correction term, it can be seen that to leading order, the amplitude A propagates with the *group velocity* c_g, while the aforementioned small correction term gives a dispersive term proportional to $t^{-1/2}$. It is important to note here that within this narrow-band approximation, the amplitude A satisfies the linear Schrödinger equation

$$i(A_t + c_g A_x) + \delta A_{xx} = 0. \tag{9}$$

Further, it can be shown that as $t \to \infty$,

$$A(x, t)|_{x=c_g t} \sim F(k_0) \left(\frac{\pi}{|\delta|t}\right)^{1/2} \exp\left(\frac{i\pi}{4}\text{sign } \delta\right), \quad \text{as } t \to \infty. \tag{10}$$

This same result can be obtained directly from (4) by using the method of stationary phase, valid here in the limit when $t \to \infty$ (see, for instance, Whitham 1974; Lighthill 1978). In this case, it is not necessary to assume also that $F(k)$ is centred at k_0, and instead k_0 is defined as a function of x/t by the relation $x = c_g(k_0)t$. Thus, in this linearized approximation, wave packets are the generic longtime outcome of the initial-value problem.

The natural extension of the group velocity to higher spatial dimensions is

$$\mathbf{c}_g = \nabla_{\mathbf{k}}\, \omega(\mathbf{k}), \tag{11}$$

obtained by differentiation of the dispersion relation (2). This result can be readily obtained by extending the above argument to several space dimensions, or perhaps more naturally by using the kinematic theory of waves (see Whitham 1974; Lighthill 1978). Thus, let the wave field be defined asymptotically by

$$\eta(\mathbf{x}, t) \sim A(\mathbf{x}, t) \exp\left(i\Theta(\mathbf{x}, t)\right) + \text{c.c.}, \tag{12}$$

where $A(\mathbf{x}, t)$ is the (complex) wave amplitude, and $\Theta(\mathbf{x}, t)$ is the phase, which is assumed to be rapidly varying compared to the amplitude. Then it is natural to define the local wave frequency and wavenumber by

$$\omega = -\Theta_t, \qquad \mathbf{k} = \nabla_\mathbf{x}\Theta. \tag{13}$$

Note that expression (10) has the required form (12). Then cross-differentiation leads to the kinematic equation for the conservation of waves,

$$\mathbf{k}_t + \nabla_\mathbf{x}\omega = 0. \tag{14}$$

But now, if we suppose that the dispersion relation (2) holds for the frequency and wavenumber defined by (13), then we readily obtain

$$\mathbf{k}_t + \mathbf{c}_\mathrm{g} \cdot \nabla\mathbf{k} = 0, \tag{15}$$

with a similar equation for the frequency. Thus, both the wavenumber and frequency propagate with the group velocity, a fact which can also be seen in (10). Equation (15) is itself a simple wave equation, which can be readily integrated by the method of characteristics, or rays. It is important to note that the group velocity c_g is a function of k, so that (15) is a nonlinear equation.

Because the group velocity is the velocity of the wave packet as a whole, it is no surprise to find that it can also be identified with energy propagation. Indeed, it can be shown that in most linearized physical systems, an equation of the following form can be derived,

$$\frac{\partial \mathbf{E}}{\partial t} + \nabla \cdot (\mathbf{c}_\mathrm{g}\mathbf{E}) = 0, \tag{16}$$

where $\mathbf{E}$ is the wave action density and is proportional to the square of the wave amplitude $|A|^2$, with the factor being a function of the wavenumber $\mathbf{k}$. Typically, the wave action is just the wave energy density divided by the frequency, at least in inertial frames of reference. For water waves, when η in (1) represents the elevation of the free surface, $E = 2\rho g|A|^2$.

For water waves, the dispersion relation (3) is isotropic, and so the group velocity is in the direction of the wavenumber $\mathbf{k}$ and has a magnitude $c_\mathrm{g} = |\mathbf{c}_\mathrm{g}|$ given by

$$c_\mathrm{g} = \partial\omega/\partial\kappa = \frac{c}{2}\left\{\frac{1 + 3Bq^2}{1 + Bq^2} + \frac{2q}{\sinh 2q}\right\}. \tag{17}$$

Here, we recall that $q = \kappa h$. Assuming that the Bond number $B < 1/3$, it can be shown that the phase speed $c = \omega/\kappa$ decreases from $c_0 = \sqrt{gh}$ when $\kappa = 0$ to a minimum of c_m at $\kappa = \kappa_\mathrm{m}$ and then increases to infinity as κ increases from κ_m to infinity. It is important to note that at the bifurcation point $\kappa = \kappa_\mathrm{m}$, $c = c_\mathrm{g} = c_\mathrm{m}$. Then gravity waves are defined by the wavenumber range $0 < \kappa < \kappa_\mathrm{m}$, $c_\mathrm{g} < c$, while capillary waves are defined by the wavenumber range $\kappa > \kappa_\mathrm{m}$, $c_\mathrm{g} > c$. In the absence of surface tension (i.e. $B \to 0$), $c_\mathrm{g} < c$ for all wavenumbers $\kappa > 0$, and in the deep water limit ($q \to \infty$) $c_\mathrm{g} \approx c/2$. When the Bond number $B > 1/3$, then the phase speed and the group velocity both increase as κ is increased from zero.

2 Weakly nonlinear waves

2.1 Nonlinear Schrödinger equation

The theory of linearized waves described in the previous section is valid when initial conditions are such that the waves have sufficiently small amplitudes. However, after a sufficiently long time (or if the initial conditions describe waves of moderate or large amplitudes), the effects of the nonlinear terms in the physical system need to be taken into account. As discussed in Chapter 1, there are two principal cases when weak nonlinearity needs to be taken into account, namely, long waves and wave packets. The first case has been discussed in the previous chapters, and it is the second case which is of interest here.

First, let us consider unidirectional waves, where, as described above, the linear theory predicts that a localized initial state will typically evolve into wave packets, with a dominant wavenumber k and corresponding frequency ω given by (2), within which each wave phase propagates with the phase speed c, but whose envelope propagates with the group velocity c_g (6, 17). Note that we have replaced k_0, ω_0 of the previous section with k, ω for notational convenience. After a long time, the packet tends to disperse around the dominant wavenumber, which tendency is opposed by cumulative nonlinear effects. In order to describe this situation, we replace the linear expression (7) with

$$\eta = \epsilon A(X, T) \exp(i\theta) + \text{c.c.} + \cdots, \tag{18}$$

where

$$\theta = kx - \omega(k)t, \quad X = \epsilon(x - c_g t), \quad T = \epsilon^2 t. \tag{19}$$

Here $\epsilon \ll 1$ is a small parameter measuring the wave amplitude and we have scaled the linear dispersive effects to balance the leading order nonlinear effects, while the omitted terms are $O(\epsilon^2)$. The outcome for these unidirectional waves is described by the NLS equation

$$iA_T + \delta A_{XX} + \mu |A|^2 A = 0. \tag{20}$$

Here, the coefficient δ of the linear dispersive term is defined by (6) and we note that the linear part of (20) agrees with (9) when we take account of the rescaling (19). The coefficient μ of the nonlinear term needs to be found separately for each specific physical system. The NLS equation is integrable in both the focusing ($\delta\mu > 0$) and the defocusing ($\delta\mu < 0$) cases.

In order to indicate how this derivation proceeds, we shall give a brief description below in Section 2.2 for the case of capillary–gravity waves. The NLS was first derived for water waves in the absence of surface tension by Zakharov (1968) for the case of deep water and then by Hasimoto and Ono (1972) for finite depth (see also the recent review by Dias and Bridges 2005). In this case the coefficient μ is given by

$$\mu = -\frac{k^2\omega}{4\sigma^4}(9\sigma^4 - 10\sigma^2 + 9) + \frac{gk\omega}{2\sigma^2(gh - c_g^2)}(2\sigma(3 - \sigma^2) + 3q(1 - \sigma^2)^2),$$

(21)

where $\sigma = \tanh q$ and $q = kh$. Note that the first term is always negative and the second term is always positive. In deep water ($q \to \infty$), the second term vanishes and the coefficient $\mu \to -2\omega k^2 < 0$. In general, $\mu < 0$ (>0) according as $q > q_c$ ($q < q_c$), where $q_c = 1.36$, and with equality between the two terms occurring at $q = q_c$.

When surface tension is present, the coefficient was obtained by Kawahara (1975) and is given by

$$\mu = -\frac{\omega k^2}{4\sigma^2(1 + Bq^2)(\sigma^2 + Bq^2(\sigma^2 - 3))}\{(9\sigma^4 - 10\sigma^2 + 9)$$
$$+ Bq^2(15\sigma^4 - 44\sigma^2 + 30) + B^2q^4(6\sigma^4 - 25\sigma^2 + 21)\}$$
$$+ \frac{gk\omega}{2\sigma^2(gh - c_g^2)}\{(1 + Bq^2)(6\sigma + 3q(1 - \sigma^2)^2) - 2(1 + 3Bq^2)\sigma^3\}. \quad (22)$$

Note the singularities when $\sigma^2(1 + Bq^2) = 3Bq^2$ defining $q = q_S$ say, where the system supports a second harmonic resonance (i.e. $c(k) = c(2k)$), also know as Wilton's ripples in this context. We recall that when $B < 1/3$, the phase speed c decreases from the value $c_0 = \sqrt{gh}$ when $q = 0$ to a minimum c_m at $q = q_m$ and then increases to infinity as q increases from q_m to infinity. It follows that $q_S < q_m$. On the other hand when $B > 1/3$, there is no second harmonic resonance. There is another singularity when $c_g^2 = gh$ defining $q = q_L$ say, due to a long-short wave resonance, which also exists only when $B < 1/3$. It is readily seen that $q_L > q_m$. The derivation of the NLS equation fails in both cases $q = q_S$, and $q = q_L$. The first term in the expression (22) now changes sign at q_S (second harmonic resonance) and the second term changes sign at q_L (long-short wave resonance). These considerations, together with an evaluation of the respective limits $q \to 0, \infty$ are sufficient to establish that when $B < 1/3$, the coefficient μ changes sign four times as q increases from zero to infinity. These occur at the critical values $q_{c1} < q_S < q_L < q_{c2}$, where $\mu > 0$ for $q < q_{c1}, q_S < q < q_L, q > q_{c2}$, and $\mu < 0$ otherwise (see Fig. 1). When $B > 1/3$, similar considerations show that μ changes sign only once at $q = q_{c3}$ say, where $\mu > 0$ ($\mu < 0$) for $q > q_{c3}$ ($q < q_{c3}$) respectively.

In deep water ($q \to \infty$), the coefficient (22) reduces to

$$\mu = -\frac{k^2\omega}{4}\frac{(8 + Wk^2 + 2W^2k^4)}{(1 + Wk^2)(1 - 2Wk^2)},$$

(23)

where $W = \Sigma/g$ now measures the effect of surface tension. Note that this limit is distinguished from the limit $q \to \infty$ with B fixed, in that we take the joint

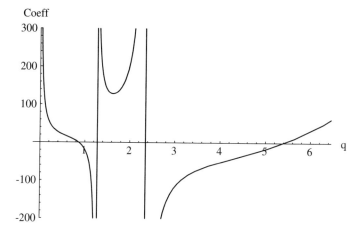

Figure 1: The nonlinear coefficient μ (22) (made non-dimensional using a length scale h and a velocity scale $\sqrt{gh}$) for Bond number $B = 0.2$.

limit $q \to \infty$ and $B \to 0$ with $W = Bh^2$, or $Wk^2 = Bq^2$ fixed. In this deep-water limit, the only change of sign occurs at the second harmonic resonance, $k = k_S$, where $2Wk_S^2 = 1$, and $\mu < 0$ ($\mu > 0$) for $k < k_S$ ($k > k_S$), respectively.

2.2 Derivation for capillary–gravity waves

The full Euler equations for an inviscid, incompressible fluid in irrotational flow can be reduced to the solution of Laplace's equation for a velocity potential $\phi = \phi(x, z, t)$,

$$\frac{\partial^2 \phi}{\partial x^2} + \frac{\partial^2 \phi}{\partial z^2} = 0, \quad -h < z < \zeta. \tag{24}$$

Here $z = \zeta(x, t)$ denotes the location of the free surface, and we will restrict our attention to the two-dimensional case when there is no dependence on the horizontal transverse variable y. This is to be solved with boundary conditions at the rigid bottom $z = -h$ and at the free surface,

$$\phi_z = 0 \quad \text{at } z = -h, \tag{25}$$

$$\zeta_t + u\eta_x = w, \quad \text{at } z = \zeta, \tag{26}$$

$$\phi_t + g\zeta + \frac{1}{2}|\mathbf{u}|^2 = \Sigma \frac{\zeta_{xx}}{(1 + \zeta_x^2)^{3/2}}, \quad \text{at } z = \zeta. \tag{27}$$

Here $\mathbf{u} = (u, w) = (\phi_x, \phi_z)$ is the velocity field.

To derive the NLS equation (20), we now seek an asymptotic expansion of the form (18),

$$\zeta = \epsilon A(X, T) \exp(i\theta) + \text{c.c.} + \epsilon^2 \eta^{(2)} + \epsilon^3 \eta^3 + \cdots, \tag{28}$$

where the phase variable θ and the slow variables X and T are defined by (19). There are analogous expansions for u, w and ϕ. At the leading order, we obtain the dispersion relation (3) with κ replaced with k. It is convenient to write this in the form (2),

$$D(\omega, k) \equiv \omega^2 - gk(1 + Bq^2)\tanh q = 0, \tag{29}$$

where we recall that $q = kh$. At the next order, we find that the envelope A moves with group velocity, a result already anticipated in Section 1 (see (5, 6) and the following discussion).

We then seek expressions for the nonlinear component $\eta^{(2)}$ in the form

$$\zeta^{(2)} = \zeta_2^{(2)}(X, T)\exp 2i\theta + \text{c.c.} + \zeta_0^{(2)}(X, T). \tag{30}$$

To leading order the second harmonic $\zeta^{(2)}$ is governed by an equation of the form

$$D(2\omega, 2k)\zeta^{(2)} = \nu_2 A^2, \tag{31}$$

where ν_1 is a constant that depends on k and B. Then provided that there is no second harmonic resonance, i.e. $D(2\omega, 2k) \neq 0$ and so $q \neq q_S$, it follows that to leading order, $\zeta_2^{(2)} = \nu_2 A^2/D(2\omega, 2k)$. The mean term $\zeta_0^{(2)}$ satisfies an equation of the form

$$\frac{\partial^2 \zeta_0^{(2)}}{\partial t^2} - gh\frac{\partial^2 \zeta_0^{(2)}}{\partial x^2} = \nu_0\,\epsilon^2|A|_{XX}^2, \tag{32}$$

where ν_0 is a constant that depends on k and B. The simplest method to obtain this equation is to average the full Euler equations over the phase θ; the left-hand side of (32) is easily recognized as the long-wave operator, as could be expected. Indeed, taking the limit $\omega, k \to 0$ in (29) gives $D(\omega, k) \approx \omega^2 - ghk^2$, which is the Fourier transform of the left-hand side of (32). Then, taking account of the fact that $\zeta_0^{(2)}$ depends on the slow variables X and T, it is readily seen that this equation can be solved at leading order by a term of the form $\zeta_0^{(2)} = \nu_0|A|^2/(c_g^2 - gh)$. Note that we must exclude the long-short wave resonance when $q = q_L$ or $c_g^2 = gh$.

It is pertinent to note here that the possible occurrence of a second harmonic resonance, or a long-short wave resonance, at certain wavenumbers is indicative of the possible presence of resonant wave triads. That is, the wave with a wavenumber–frequency pair (k, ω) may be in resonance with the pairs (k_1, ω_1) and (k_2, ω_2), i.e. whenever $k_1 + k_2 = k$, $\omega_1 + \omega_2 = \omega$. Such triads exist for capillary–gravity waves when $0 < B < 1/3$. The presence of a resonant triad will allow a modulation of the primary wave on a time-scale of ϵ^{-1}, which is an order of magnitude faster than that being considered here. Thus, in this circumstance, for the validity of the NLS model, it is necessary to suppose a priori that the wavenumber spectrum is sufficiently narrow-banded to exclude any resonant triads.

At the third order in ϵ, the term of interest in $\zeta^{(3)}$ is that proportional to $\exp i\theta$, say $\zeta_1^{(3)}\exp i\theta$. At leading order, we get an equation of the form

$$D(\omega, k)\zeta_1^{(3)} = 2\omega(iA_T + \delta A_{XX}) + i\mu_2\zeta_2^{(2)}A^* + i\mu_0\zeta_0^{(2)}A + i\mu_1 A^2 A^*. \tag{33}$$

Here μ_0, μ_1 and μ_2 are constants depending on k and B, and A^* is the complex conjugate of A. The origin of the linear terms on the right-hand side of (33) has been discussed in Section 1 (see (5, 6) and the following discussion). The first two nonlinear terms arise due to the quadratic interaction of the second harmonic and the mean with the primary harmonic, while the third nonlinear term is due to the cubic self-interaction of the primary harmonic. Since the dispersion relation (29) holds, the left-hand side of (33) is zero, and so the NLS equation (20) is obtained with

$$2\omega\mu = \frac{\mu_2 v_2}{D(2\omega, 2k)} + \frac{\mu_0 v_0}{c_g^2 - gh} + \mu_1. \tag{34}$$

3 Envelope solitons

The NLS equation (20) is a canonical model for weakly nonlinear wave packets in many physical systems. For instance, in the fluid mechanics context it has been derived for internal waves by Grimshaw (1977a), for coastally trapped waves by Grimshaw (1977b), and for equatorial waves by Boyd (1983). In a slightly different context the NLS equation also governs waves on thin vortex filaments (Hasimoto 1972).

Like the Korteweg–de Vries equation, the NLS equation is integrable with an associated inverse scattering transform, a result first shown by Zakharov and Shabat (1972). There are two cases, the so-called focusing NLS equation when $\delta\mu > 0$ and the defocusing NLS equation when $\delta\mu < 0$. For water waves without surface tension $2\delta = \partial c_g / \partial k < 0$, and so we have the focusing (defocusing) NLS equation according as $\mu < 0 \, (> 0)$, i.e. as above from (21), as $q \, (= kh) > 1.36 \, (q < 1.36)$.

When surface tension is included, both δ and μ can change sign. For instance, in deep water, $\delta < 0 \, (> 0)$ according as $k < k_d \, (k > k_d)$ where the critical value is defined by $W k_d^2 = 0.155$; we note for later use that for these deep-water capillary-gravity waves, $c > c_g \, (c < c_g)$ for $k < k_m \, (k > k_m)$, where the bifurcation value is here defined as $c = c_g$, given by $W k_m^2 = 1$. Thus at the bifurcation point, $\delta > 0$ since $k_m > k_d$. On the other hand, as above from (23), the nonlinear coefficient $\mu < 0 \, (> 0)$ according as $k < k_S \, (k > k_S)$, recalling that $W k_S^2 = 0.5$; note that unlike the case for δ, $\mu \to \infty$ at this transition point. Since we have $k_d < k_S < k_m$, it follows that the NLS equation (20) is of the focusing kind for $k < k_d$ and for $k > k_S$, and is otherwise defocusing. In particular, at the bifurcation point where $c = c_g$, $\mu, \delta > 0$, the NLS equation is of the focussing kind.

In the general case for capillary–gravity waves, let us first consider $0 < B < 1/3$. Then $\delta < 0 \, (> 0)$ according as $q = kh < q_d \, (> q_d)$ where the critical point q_d is defined by $\delta = 0$, i.e. $\partial c_g / \partial k = 0$. While it is readily shown that $q_d < q_m$, where we recall that q_m is the bifurcation point where $c = c_g$, the location of q_d vis-á-vis q_{c1}, q_S depends on B. For instance, in the deep-water limit (here taken as $B \to 0$) $q_d < q_S$ as above, while in the limit $B \to 1/3$ it is readily shown that $q_d > q_S$. Indeed, it can be shown that $q_d < q_S$ for $0 < B < B_c$, where $B_c = 0.197$, and that $q_d > q_S$ for $1/3 > B > B_c$. In this latter case, we can now infer that the NLS equation (20) is of the focusing kind for $q_{c1} < q < q_S, q_d < q < q_L, q > q_{c2}$, and

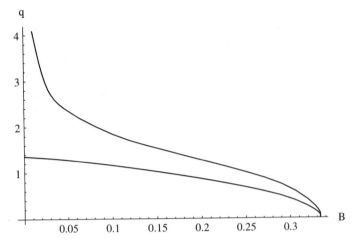

Figure 2: Plots of q_d (upper curve) and q_{c1} (lower curve) as functions of the Bond
number B.

is otherwise defocusing. In the former case $(0 < B < B_c)$, it is readily found by
numerical evaluation of μ, δ that both q_{c1} and q_d decrease as B increases, and that
$0 < q_{c1} < q_d$ (see Fig. 2). Hence now the NLS equation (20) is of the focusing kind
for $q_{c1} < q < q_d, q_s < q < q_L, q > q_{c2}$, and is otherwise defocusing. Next, when
$B > 1/3, \delta > 0$ for all q, while $\mu > 0$ ($\mu < 0$) for $q > q_{c3}$ ($q < q_{c3}$), respectively.
Thus, now the NLS equation (20) is of the focusing kind for $q > q_{c3}$ and is otherwise
defocusing.

The focusing NLS equation has solitary wave solutions (bright solitons),
given by

$$A = a\,\mathrm{sech}(\gamma X)\exp(-i\sigma T), \quad \text{where } \mu a^2 = 2\delta\gamma^2, \ \sigma = -\frac{1}{2}\mu a^2. \tag{35}$$

Note that this one-parameter family can be extended to a two-parameter family
through the gauge transformation

$$X \to X + VT, \quad A \to A\exp(iPX + i\delta P^2 T) \quad \text{where } V = 2\delta P. \tag{36}$$

Of course, this gauge transformation amounts to a small shift $k \to k + \epsilon P$ in the
carrier wavenumber, with consequent small shifts in the carrier frequency and group
velocity. On the other hand, the defocusing NLS equation has no such solitary wave
solutions that decay to zero at infinity; instead it has solitary waves riding on a
non-zero background (dark solitons).

A key property of the NLS equations is that plane waves are modulationally
unstable (stable) in the focusing (defocusing) case. That is, the NLS equation has
the exact plane wave solution,

$$A = A_0\exp(i\mu|A_0|^2 t), \tag{37}$$

which is then perturbed with a small-amplitude modulation proportional to $\exp{(iKx - \Omega t)}$. It is readily found that the growth rate Ω is given by

$$\Omega^2 = K^2(2\delta\mu|A_0|^2 - \delta^2 K^2). \tag{38}$$

Thus in the focusing NLS case when $\delta\mu > 0$, there is a positive growth rate for modulation wavenumbers K such that $K^2 < 2\mu|A_0|^2/\delta$. On the other hand, Ω is pure imaginary for all K in the defocusing case when $\delta\mu < 0$. The maximum growth rate occurs for $K = K_M = (\mu/\delta)^{1/2}|A_0|$ and the instability is due to the generation of side bands with wavenumbers $k \pm K_M$. As the instability grows, the full NLS equation (20) is needed to describe the longtime outcome of the collapse of the uniform plane wave into several soliton wave packets, each described by (35).

The implication for water waves (i.e. the Bond number $B = 0$) is that plane Stokes waves in deep water ($q = kh > 1.36$) are unstable. This remarkable result was first discovered by Benjamin and Feir in 1967 using a different theoretical approach, again by Zakharov (1968) from the NLS equation, and has since been confirmed by experiments. In the presence of surface tension, for the case $0 < B < 1/3$, plane waves are modulationally unstable in the wavenumber bands $q_{c1} < q < q_d, q_S < q < q_L, q > q_{c2}$, while if $B > 1/3$, there is modulational instability for $q > q_{c3}$.

4 Bifurcation to solitary waves

Although the soliton (35) is an important solution of the the NLS equation (20), it does not describe a solitary wave of the full system, since, in general, the phase propagates at a speed c and the envelope with a different speed c_g (see (18)). But when the carrier wavenumber k is such that $c = c_g$ (i.e. $q = q_m$), the phase and envelope propagate at the same speed and we see that a true solitary wave is then possible. However, the soliton solution (35) introduces a correction of $\epsilon^2\sigma/k$ to the phase speed c, and hence it is necessary to calculate a similar higher-order correction to the group velocity. It transpires that a gauge transformation by itself is not sufficient for this purpose, and instead we must introduce higher-order terms into the NLS equation (20).

When generic higher-order terms of relative $O(\epsilon)$ are added to (20), we get

$$iA_T + \delta A_{XX} + \mu|A|^2 A + i\epsilon\{\delta_1 A_{XXX} + \mu_1|A|^2 A_X + \mu_2 A^2 A_X^*\} = 0, \tag{39}$$

where

$$\delta_1 = -\frac{1}{6}\frac{\partial^3\omega}{\partial k^3}. \tag{40}$$

While the higher-order nonlinear coefficients μ_1 and μ_2 can be expressed as derivatives with respect to k of the nonlinear coefficient μ, their exact values are not required here. An envelope solitary wave is now given by

$$A = \{R(X - VT) + i\epsilon R_1(X - VT)\}\exp{(i(PX - \sigma T) + i\theta_0)}. \tag{41}$$

Without any loss of generality, we can assume that both R and R_1 are real-valued, as any imaginary part to R_1 can be absorbed into R and the phase constant θ_0 accounts for a phase in the amplitude. Note that it is also necessary to allow for a wavenumber shift from k to $k + \epsilon P$ here. Substitution into (39) now yields

$$(\delta - 3\epsilon\delta_1 P)R_{XX} + (\sigma - \delta P^2 + \epsilon\delta_1 P^3)R + (\mu - \epsilon(\mu_1 + \mu_2)P)R^3 = 0, \quad (42)$$

$$\delta_1 R_{1XX} + (\sigma - \delta P^2)R_1 + \mu R^2 R_1 + F_1 = 0, \quad (43)$$

where

$$F_1 = \delta_1 R_{XXX} - (V_1 + 3\delta_1 P^2)R_X + (\mu_1 + \mu_2)R^2 R_X. \quad (44)$$

Here we have expanded $V = 2\delta P + \epsilon V_1$. At leading order in ϵ, we obtain the previous result that R and V are given by (35), when the gauge transformation (36) is accounted for. Next, we see that the extended equation (42) for R is identical in form to the leading order equation, and hence has the same solution, but with modified coefficients. That is, δ is replaced by $\delta - 3\epsilon\delta - 1P$, σ is replaced by $\sigma + \epsilon\delta_1 P^3$ and μ is replaced by $\mu - \epsilon(\mu - 1 + \mu_2)$. The inhomogeneous equation (43) for R_1 needs a compatibility condition, i.e. F_1 should be orthogonal to R, which is the bounded solution of the homogeneous equation for R_1 (i.e. (43) with $F_1 = 0$). Thus we require that

$$\int_{-\infty}^{\infty} RF_1 dX = 0. \quad (45)$$

But from (44), this is automatically satisfied and imposes no constraint on R_1. But if we look further into (43), we see that as $X \to \pm\infty$, F_1 has a leading order term proportional to $\exp(\mp\gamma X)$. Since the bounded solution to the homogeneous equation for R_1 is just R, it will have the same behaviour. Hence, the inhomogeneous equation will have solutions behaving like $X \exp(\mp\gamma X)$ as $X \to \pm\infty$, which is unacceptable. To avoid this, we must put to zero the coefficient of $\exp(\mp\gamma X)$ in F_1 as $X \to \pm\infty$, which gives

$$V_1 = -3\delta_1\kappa^2 \quad \text{so that} \quad V = 2\kappa\delta_1 - \epsilon3\delta_1\kappa^2. \quad (46)$$

It now follows that the total phase speed is

$$c = \frac{\omega(k) + \epsilon Pc_g(k) + \epsilon^2\sigma}{k + \epsilon P}, \quad (47)$$

or

$$c = \frac{\omega(k + \epsilon P)}{k + \epsilon P} - \frac{\epsilon^2\mu a^2}{2k} + O(\epsilon^3). \quad (48)$$

That is, the linear phase speed $c_p(k + \epsilon P)$ slightly modified by finite amplitude effects. On the other hand, the envelope speed is

$$c_g(k) + \epsilon V = c_g(k + \epsilon P) + O(\epsilon^3), \quad (49)$$

after using (46), which is just the linear group velocity to leading order. Since these are not equal, in general, this envelope soliton is not a true solitary wave of the full equations.

But now, following the argument of Akylas (1993) of the case for deep-water capillary–gravity waves (extended to finite depth by Maleewong *et al.* 2005), suppose that the linear phase speed has a local extremum at a particular value of the wavenumber $k = k_m$. Then the linear group velocity $c_g(k_m)$ is equal to the linear phase speed $c(k_m)$. It follows that the bifurcation condition for a small-amplitude envelope solitary wave can be found by equating (48) with (49), assuming $k + \epsilon P = k_m + \epsilon^2 \Delta$ (i.e. in effect $k = k_m$ and $P = \epsilon \Delta$). This gives

$$\Delta = -\frac{\mu}{\delta} \frac{a^2}{4k_m}. \tag{50}$$

Since we must have $\delta \mu > 0$ at $k = k_m$, it follows that $\Delta < 0$ and so $k < k_m$ in all cases. Let us now note that when $\delta > 0$ (<0), the phase speed c has a minimum (maximum) at $k = k_m$ since it is readily shown that $kc_{kk} = c_{gk} = 2\delta$ at $k = k_m$. Thus, in the case when $\mu > 0$, $\delta > 0$, the phase speed c (48) of the solitary wave is less than the linear phase speed $c(k_m)$ as required, while in the opposite case $\mu < 0$, $\delta < 0$, the phase speed c (48) is of the solitary wave is greater than the linear phase speed $c(k_m)$ as required. A plot of the leading order term in the solution (41) is shown in Fig. 3 in scaled coordinates for a case when the ratio of the carrier wavenumber to the envelope wavenumber is 5, i.e. $(k + \epsilon P)/\epsilon \gamma = 5$.

It is pertinent to note that the phase θ_0 in the solution (41) is not determined by this asymptotic analysis and is apparently free. It is equal to 0 in Fig. 3, which can thus be called a wave of elevation. The choice $\theta_0 = \pi$ would produce an inverted shape, which is a wave of depression. However, a delicate and rigorous analysis of

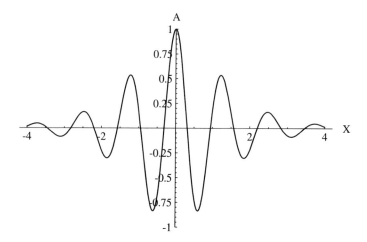

Figure 3: Plot of an envelope solitary wave, in scaled coordinates, for a case when the ratio of the carrier wavenumber to the envelope wavenumber is 5.

the full equations for the physical system in question reveals that in fact the phase θ_0 is not arbitrary, and only these two possibilities can in general persist in the full system (see Iooss and Pérouème 1993; Dias and Kharif 1999; Grimshaw and Iooss 2003; Dias and Iooss 2003; Dias and Bridges 2005). The only permitted choices are $\theta_0 = 0, \pi$, so that these solitary waves are symmetric and are either waves of elevation or depression. Further, based on an analysis by Calvo $et\ al.$ (2000) for a certain model system, it would seem that the elevation wave is unstable, while the depression wave is stable. However, the growth rate of the instability may be very small in practice.

From our discussion above in Section 3 for capillary–gravity waves, we see that indeed both the nonlinear coefficient $\mu > 0$ and the dispersive coefficient $\delta > 0$ at the bifurcation point $k = k_m$, where the linear phase speed is a minimum, and $c = c_g$. Thus, this asymptotic argument implies the existence of true envelope solitary waves when $0 < B < 1/3$. Some experimental evidence for the existence of capillary–gravity solitary waves has been reported by Longuet-Higgins and Zhang (1997) and Zhang (1995). Numerical simulations of the full Euler equations have been reported by Longuet-Higgins (1989) and Vanden-Broeck and Dias (1992) for deep water, and by Dias $et\ al.$ (1996) and Maleewong $et\ al.$ (2005) for the finite depth case. A typical numerical solution is shown in Fig. 4. In general, for an elevation wave, as the wave amplitude increases, the crests and troughs separate, and the crests fall in amplitude while the troughs deepen; at a sufficiently large amplitude, the central crest becomes negative, and the wave is a combination of two large troughs, see Fig. 5. In contrast, for a depression wave, as the amplitude increases, the outlying crests and troughs diminish, leaving just a central trough; eventually a limiting configuration is reached, which contains a trapped bubble.

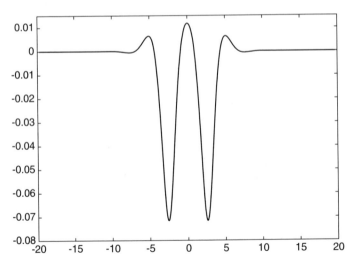

Figure 4: Plot of an envelope solitary wave for Bond number $B = 0.27$ and Froude number $F = c/\sqrt{gh} = 0.9486$.

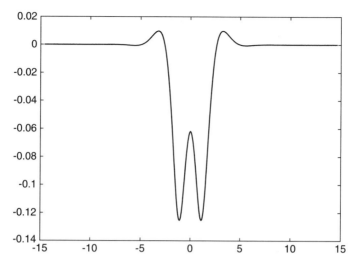

Figure 5: Plot of an envelope solitary wave for Bond number $B = 0.25$ and Froude number $F = c/\sqrt{gh} = 0.9165$.

A rigorous existence proof can be obtained from an approach based on the techniques of spatial dynamics (see Iooss and Kirchgassner 1990; Iooss and Pérouème 1993; Dias and Iooss 1993; 2003; Groves 2004; Dias and Bridges 2005). In the deepwater limit, a curious new feature emerges, in that although the envelope soliton (35) decays exponentially to zero as $X \to \pm\infty$, there is a higher-order mean-flow term which only decays to zero algebraically (as $1/X^2$) (see Iooss and Kirrmann 1996; Akylas *et al.* 1998). For the deep-water case, the stability to perturbations in the propagation direction was investigated by Calvo and Akylas (2002) numerically. They found that the depression wave is stable at all amplitudes; however, the elevation wave is unstable at small amplitudes, but stabilizes at some finite amplitude where the wave profile features two well-separated troughs. However, recently Kim and Akylas (2006b) have shown that the depression wave is unstable to transverse perturbations.

The arguments given above suggest that such envelope solitary waves can arise in any physical system where the necessary bifurcation takes place (i.e. $c = c_g$), and where the coefficients μ and δ in the NLS equation (20) have the same sign; some preliminary results in this direction have been given by Grimshaw and Iooss (2003). For instance, a study of interfacial waves in this context has been reported by Calvo and Akylas (2003) (who also considered the stability issue with an outcome similar to that reported above for small-amplitude waves), Dias and Iooss (1996, 2003) and Iooss (1999).

A new feature that emerges at finite amplitude is the presence of so-called multi-bump solitary waves. These consist of the superposition of two or more envelope solitary waves, as described above, connected by an appropriate phase matching of their decaying tails. They have been studied in certain model systems by Champneys and Toland (1993) and Buffoni *et al.* (1995), and in the full Euler equations by Yang and Akylas (1997) and Buffoni and Groves (1999). Some numerical solutions have been exhibited by Dias *et al.* (1996) and Maleewong *et al.* (2005). A physical explanation for their existence can be found by considering the NLS solitary wave (41). To this point, we have considered only symmetric solitary waves for which the phase constant $\theta_0 = 0, \pi$. Let us now suppose we choose $0 < \theta_0 < \pi$, so that the envelope is not symmetric. This is permissible at leading order when the solitary wave satisfies the NLS equation (20), but as discussed above, such solutions do not persist at higher order. Instead, as shown by Yang and Akylas (1997), growing oscillations appear on one side of the wave packet, with exponentially small amplitude. But these growing oscillations eventually form a new wave packet, and for certain values of the phase θ_0 the disturbance then terminates leading to an envelope solitary wave with two bumps. Otherwise, the process continues and a third wave packet forms, and so on. The outcome is a countable infinity of such multi-bump waves, each bifurcating at a finite amplitude.

5 Two-dimensional solitary waves

When the effects of modulation in the transverse y-direction are taken into account, the wave amplitude is given by $A(X, Y, T)$, where $Y = \epsilon y$ (see (18)). The NLS equation for capillary–gravity waves is replaced by the Benney–Roskes system (Benney and Roskes 1969), also widely known as the Davey–Stewartson equations (Davey and Stewartson 1974),

$$iA_T + \delta A_{XX} + \delta_1 A_{YY} + \mu |A|^2 A + UA = 0, \tag{51}$$

$$\alpha U_{XX} + U_{YY} + \beta(|A|^2)_{YY} = 0, \tag{52}$$

where

$$\delta_1 = \frac{c_g}{2k}, \quad \text{and} \quad \alpha = 1 - \frac{c_g^2}{gh}. \tag{53}$$

Here, U is an induced mean flow term. The coefficients μ and δ are those defined in (20); for simplicity, we present the coefficient β in the absence of surface tension,

$$(gh - c_g^2)\beta = -\frac{2\omega^3}{\sigma^2} \left(1 + \frac{c_g}{2c}(1 - \sigma^2)\right)^2. \tag{54}$$

For the case when the surface tension effects are included, see Djordjevic and Redekopp (1977). The system (51, 52) is not integrable in general, but can be shown to reduce to an integrable Davey–Stewartson system in the limit $q \to 0$, either DSII if $0 \leq B < 1/3$, or DSI if $B > 1/3$.

For the case when there is no surface tension, $\delta < 0, \delta_1 > 0, \beta > 0$ and $\alpha > 0$ so that the equation for A is 'hyperbolic' , but that for U is 'elliptic'. Also, when we recall that $\mu < 0 (> 0)$ according as $q > q_c (< q_c), q_c = 1.36$, we see that the equation for A is focusing for 'x' and defocusing for 'y' when $q > q_c$, but is defocusing for 'x' and focusing for 'y' when $q < q_c$. The variable U that appears here is a wave-induced mean flow, which tends to 0 in the limit of deep-water waves, $q = kh \to \infty$.

The system (51, 52) again has the plane wave solution (37), whose modulational stability can be analysed in a manner similar to that described above in the context of the NLS equation. Thus, the plane wave (37) is now perturbed with a small amplitude modulation proportional to $\exp(iKX + iLY - \Omega T)$. The growth rate is now given by

$$\Omega^2 = (\delta K^2 + \delta_1 L^2) \left(\left\{ \mu + \frac{\beta L^2}{\alpha K^2 + L^2} \right\} 2|A_0|^2 - \delta K^2 - \delta_1 L^2 \right). \tag{55}$$

In the absence of surface tension, the coefficients $\delta < 0, \delta_1 > 0, \alpha > 0, \beta > 0$ and $\mu > 0 \, (< 0)$ according as $q < q_c \, (> q_c)$. The outcome is that now instability can occur for all values of q and exists in a band in the K–L plane where K and L are the modulation wavenumbers. The instability is purely two-dimensional when $q < q_c$, and the band becomes narrower and the growth rate weaker as $q \to 0$; curiously, the instability band reduces to a line in the K–L plane when $q = 0.38$. For more details, see Benney and Roskes (1969), Mei (1983), Dias and Kharif (1999) and Dias and Bridges (2005).

When surface tension is included, the situation becomes more complicated. When the Bond number $B > 1/3$, the coefficients $\delta > 0, \delta_1 > 0, \alpha < 0$ for all q, while the coefficient $\mu > 0 \, (\mu < 0)$ for $q > q_{c3} \, (q < q_{c3})$. Thus, the coupled system is elliptic–hyperbolic for all q and is focusing with respect to 'x' and defocusing with respect to 'y' when $q > q_{c3}$, and is focusing with respect to 'y' and defocusing with respect to 'x' when $q < q_{c3}$. On the other hand when the Bond number B is such that $0 < B < 1/3$, the coefficient $\delta_1 > 0$ for all q, but $\delta < 0 \, (> 0)$ according as $q < q_d \, (> q_d)$ where we recall that q_d is defined by the condition that $\partial c_g / \partial k = 0$. Also now, $\alpha > (< 0)$ according as $q < q_L \, (> q_L)$ where we recall that $q_L \, (> q_d)$ is defined by the condition that $c_g^2 = gh$. It follows that for gravity waves, defined here by $q < q_d$, we have the case that the coupled system is hyperbolic–elliptic. However, in the regime for capillary waves, defined here by $q > q_L$, the coupled system is elliptic–hyperbolic. In the intermediate regime for gravity–capillary waves, $q_d < q < q_L$, the coupled system is elliptic–elliptic. Also, we recall that the nonlinear coefficient μ (22) changes sign four times, at the critical values $q = q_{c1}, q_S, q_L, q_{c2}$ where $q_{c1} < q_S < q_L < q_{c2}$; also $q_d < q_S \, (> q_S)$ according as $0 < B < B_c \, (B_c < B < 1/3)$ where $B_c = 0.197$, and $q_{c1} < q_d$. Hence, there are several possible combinations of the sign of the three coefficients δ, δ_1 and μ, implying that the equation for A may be focussing with respect to both variables 'x' and 'y', or focusing with respect to one of the variables 'x' and 'y' and defocusing with respect to the other variable, or defocusing with respect to both variables. Thus overall, there are several possible equation types for the coupled system (51, 52). However, it will transpire that our main interest centres on the case when there is a bifurcation, namely at $q = q_m$ defined by $c = c_g$. In this case, $\delta > 0, \alpha > 0, \mu > 0$ and so the system (51, 52) is elliptic–elliptic and focusing for both variables.

The modulational stability of the plane Stokes wave (37) in the presence of surface tension is again given by (55), which defines the regions in the K–L plane where instability can occur (see, for instance, Djordjevic and Redekopp 1977).

Note that whenever both μ and $\beta < 0$ while both δ and $\delta_1 > 0$, there is no instability for any K and L. For instance, in deep water $\beta \to 0$, and it is readily shown that for $q_d < q < q_s$, there is no instability as then $\mu < 0$, $\delta > 0$, $\delta_1 > 0$. But, if $0 < q < q_d$, $\mu < 0$, $\delta < 0$, $\delta_1 > 0$ and instability occurs in a band in the K–L plane defined by $0 > \delta K^2 + \delta_1 L^2 > 2\mu|A_0|^2$, analogous to the instability which arises in the case of zero surface tension. On the other hand, if $q > q_s$, then $\mu > 0$, $\delta > 0$, $\delta_1 > 0$ and instability arises for long-wave modulations in the elliptical region defined by $\delta K^2 + \delta_1 L^2 < 2\mu|A_0|^2$.

The system (51, 52) supports the one-dimensional NLS solitary wave (35) as a special solution when there is no Y-dependence. Further, as discussed in Section 4, when $0 < B < 1/3$ there is a bifurcation from $c = c_g$ at $q = q_m$ to a steady one-dimensional solitary wave solution of the full system. Recently, it has been shown by Kim and Akylas (2005) that there is a two-dimensional counterpart to this; that is, there is a fully two-dimensional solitary wave (often called a 'lump' solution), which is localized in both the X- and Y-directions, and which also bifurcates from the wavenumber q_m. To leading order it satisfies the Benney–Roskes system (51, 52), which, as we noted above, is elliptic–elliptic and focusing in both variables at the bifurcation point. Indeed, it is essential that the system be elliptic–elliptic to ensure that there is full localization. Although the wave amplitude A has an exponentially decaying oscillatory tail, analogous to its one-dimensional counterpart, the full solitary wave decays only algebraically as $1/X^2$ due to the induced mean flow term U. In the case of finite-depth, these 'lump' solutions of (51, 52) are found numerically, although in the limit of deep water when the system reduces to a two-dimensional nonlinear Schrödinger at leading order, some rigorous existence theorems are available (see, for instance, Strauss 1977).

Numerical evidence for the existence of two-dimensional solitary waves in the full Euler equations has been reported by Parau et al. (2005). Milewski (2005), using a model system obtained from the full Euler equations by retaining only quadratic nonlinearity, also found numerical evidence for the existence of such waves. A rigorous existence theorem has been reported by Groves and Sun (2006). The latter two works base their studies on the shallow-water limit where the appropriate model equation for weakly nonlinear waves is the Kadomtsev–Petviashvili (KP) equation; for Bond numbers $B > 1/3$ this is the KPI equation which is known to support lump solutions (see Berger and Milewski 2000) for a study of the connection between the KPI equation and shallow-water waves in this context. Similar 'lump' solutions have been found in a model equation describing certain interfacial waves in the presence of surface tension by Kim and Akylas (2006a), which suggests that they may arise under similar circumstances in other physical systems as well.

References

Akylas, T.R., Envelope solitons with stationary crests. *Phys. Fluids*, **5**, pp. 789–791, 1993.

Akylas, T., Dias, F. & Grimshaw, R., The effect of the induced mean flow on solitary waves in deep water. *J. Fluid Mech.*, **355**, pp. 317–328, 1998.

Benjamin, T.B. & Feir, J.E., The disintegration of wave trains on deep water. *J. Fluid Mech.*, **27**, pp. 417–430, 1967.

Benney, D.J. & Roskes, G., Wave instabilities. *Stud. Appl. Math.*, **48**, pp. 377–385, 1969.

Berger, K. & Milewski, P.A., The generation and evolution of lump solitary waves in surface-tension-dominated flows. *SIAM J. Appl. Math.*, **61**, pp. 731–750, 2000.

Boyd, J.P., Equatorial solitary waves. Part 2: Envelope solitons. *J. Phys. Ocean.*, **13**, pp. 428–449, 1983.

Buffoni, B. & Groves, M.D., A multiplicity result for solitary gravity-capillary waves in deep water via critical-point theory. *Arch. Rat. Mech. Anal.*, **146**, pp. 183–220, 1999.

Buffoni, B., Champneys, A. & Toland, J., Bifurcation and coalescence of a plethora of homoclinic orbits for a hamiltonian system. *J. Dynamics Diff. Equ.*, **8**, pp. 221–281, 1995.

Calvo, D.C. & Akylas, T.R., Stability of steep gravity-capillary solitary waves in deep water. *J. Fluid Mech.*, **452**, pp. 123–143, 2002.

Calvo, D.C. & Akylas, T.R., On interfacial gravity-capillary solitary waves of the Benjamin type and their stability. *Phys. Fluids*, **15**, pp. 1261–1270, 2003.

Calvo, D.C., Yang, T-S. & Akylas, T.R., On the stability of solitary waves with decaying oscillatory tails. *Proc. Roy. Soc.*, **456**, pp. 469–487, 2000.

Champneys, A.R. & Toland, J., Bifurcation of a plethora of multi-modal homoclinic orbits for autonomous Hamiltonian systems. *Nonlinearity*, **6**, pp. 665–721, 1993.

Davey, A. & Stewartson, K., On three-dimensional packets of surface waves. *Proc. Roy. Soc.*, **A338**, pp. 101–110, 1974.

Dias, F. & Bridges, T., Weakly nonlinear wave packets and the nonlinear Schrodinger equation (Chapter 2). *Nonlinear Waves in Fluids: Recent Advances and Modern Applications*, CISM Courses and Lectures No. 483, ed. R. Grimshaw, Springer: Wien and New York, pp. 29–67, 2005.

Dias, F. & Iooss, G., Capillary-gravity solitary waves with damped oscillations. *Physica D*, **65**, pp. 399–423, 1993.

Dias, F. & Iooss, G., Capillary-gravity interfacial waves in infinite depth. *Eur. J. Mech. B/Fluids*, **15**, pp. 367–393, 1996.

Dias, F. & Iooss, G., Water waves as a spatial dynamical system (Chapter 10). *Handbook of Mathematical Fluid Dynamics*, eds. S. Friedlander & D. Serre, Elsevier, (North Holland), pp. 443–499, 2003.

Dias, F. & Kharif, C., Nonlinear gravity and capillary-gravity waves. *Ann. Rev. Fluid Mech.*, **31**, pp. 301–346, 1999.

Dias, F., Menasce, D. & Vanden-Broeck, J-M., Numerical study of capillary-gravity solitary waves. *Eur. J. Mech., B/Fluids*, **15**, pp. 17–36, 1996.

Djordjevic, V.D. & Redekopp, L.G., On two-dimensional packets of capillary-gravity waves. *J. Fluid Mech.*, **79**, pp. 703–714, 1977.

Grimshaw, R., The stability of continental shelf waves, I. Side-band instability and long wave resonance. *J. Aust. Math. Soc. Ser. B*, **20**, pp. 13–30, 1977a.

Grimshaw, R.H.J., The modulation of an internal gravity-wave packet, and the resonance with the mean motion. *Stud. Appl. Math.*, **56**, pp. 241–266, 1977b.

Groves, M.D., Steady water waves. *J. Nonlinear Math. Phys.*, **11**, pp. 435–460, 2004.

Groves, M.D. & Sun, S.M., Fully localised solitary-wave solutions of the three-dimensional gravity-capillary water-wave problem. *Arch. Rat. Mech. Anal.*, 2006 (submitted).

Hasimoto, H., A soliton on a vortex filament. *J. Fluid Mech.*, **51**, pp. 477–485, 1972.

Hasimoto, H. & Ono, H., Nonlinear modulation of gravity waves. *J. Phys. Soc. Japan*, **33**, pp. 805–811, 1972.

Iooss, G., Gravity and capillary-gravity periodic travelling waves for two super-posed fluid layers, one being of infinite depth. *J. Math. Fluid Mech.*, **1**, pp. 24–61, 1999.

Iooss, G. & Kirchgassner, K., Bifurcation d'ondes solitaires en presence d'une faible tension superficielle. *C.R. Acad. Sci, Paris*, **311**, pp. 265–268, 1990.

Iooss, G. & Kirrmann, P., Capillary gravity waves on the free surface of an inviscid fluid of infinite depth. Existence of solitary waves. *Arch. Rat. Mech. Anal.*, **136**, pp. 1–19, 1996.

Iooss, G. & Pérouème, M.C., Perturbed homoclinic solutions in reversible 1:1 resonance fields. *Journal of Differential Equations*, **102**, pp. 62–88, 1993.

Kawahara, T., Nonlinear self-modulation of capillary-gravity waves on liquid layer. *J. Phys. Soc. Japan*, **38**, pp. 265–270, 1975.

Kim, B. & Akylas, T.R., On gravity-capillary lumps. *J. Fluid Mech.*, **540**, pp. 337–351, 2005.

Kim, B. & Akylas, T.R., On gravity-capillary lumps. Part 2: Two-dimensional Benjamin equation. *J. Fluid Mech.*, **557**, pp. 237–256, 2006a.

Kim, B. & Akylas, T.R., Transverse instability of gravity-capillary solitary waves. *J. Eng. Maths.*, 2006b (to appear).

Lamb, H., *Hydrodynamics*, Cambridge University Press: Cambridge, 1932.

Lighthill, M.J., *Waves in Fluids*, Cambridge University Press: Cambridge, 1978.

Longuet-Higgins, M.S., Capillary-gravity waves of solitary type on deep water. *J. Fluid Mech.*, **200** , pp. 451–470, 1989.

Longuet-Higgins, M.S. & Zhang, X., Experiments on capillary-gravity waves of solitary type on deep water. *Phys. Fluids*, **9**, pp. 1963–1968, 1997.

Maleewong, M., Grimshaw, R. & Asavanant, J., Free surface flow under gravity and surface tension due to an applied pressure distribution. II Bond number less than one-third. *European J. Mechanics B/Fluids*, **24**, pp. 502–521, 2005.

Mei, C.C., *The Applied Dynamics of Ocean Surface Waves*, Wiley: New York, 1983.

Milewski, P., Three-dimensional localized solitary gravity-capillary waves. *Comm. Math. Sci.*, **3**, pp. 89–99, 2005.

Parau, E.I., Vanden-Broeck, J-M. and Cooker, M.J., Nonlinear three-dimensional gravity-capillary solitary waves. *J. Fluid Mech.*, **536**, pp. 99–105, 2005.

Strauss, W., Existence of solitary waves in higher dimensions. *Comm. Math. Phys.*, **55**, pp. 149–162, 1977.

Vanden-Broeck, J-M. & Dias, F., Gravity-capillary solitary waves in water of infinite depth and related free surface flows. *J. Fluid Mech.*, **240** , pp. 549–557, 1992.

Whitham, G.B., *Linear and Nonlinear Waves*, Wiley: New York, 1974.

Yang, T.S. & Akylas, T.R., On asymmetric gravity-capillary solitary waves. *J. Fluid Mech.*, **330**, pp. 215–232, 1997.

Zakharov, V.E., Stability of periodic waves of finite amplitude on the surface of a deep fluid. *J. Appl. Mech. Tech. Phys*, **2**, pp. 190–194, 1968.

Zakharov, V.E. & Shabat, A.B., Exact theory of two-dimensional self-focusing and one-dimensional self-modulation of waves in nonlinear media. *Soviet Physics, JETP*, **34**, pp. 62–69, 1972.

Zhang, X., Capillary-gravity and capillary waves generated in a wind wave tank: observations and theories. *J. Fluid Mech.*, **289**, pp. 51–82, 1995.

Index

Computational Methods in Multiphase Flow IV

Edited by: A. MAMMOLI, The University of New Mexico, USA and C.A. BREBBIA, Wessex Institute of Technology, UK

Fluid Dynamics is one of the most important topics of applied mathematics and physics. Together with complex flows and turbulence, multiphase flows remains one of the most challenging areas of computational mechanics, and even seemingly simple problems remain unsolved to date. Multiphase flows are found in all areas of technology, at all length scales and flow regimes. The fluids involved can be compressible or incompressible, linear or nonlinear. Because of the complexity of the problem, it is often essential to utilize advanced computational and experimental methods to solve the complex equations that describe them. Challenges in these simulations include nonlinear fluids, treating drop breakup and coalescence, characterizing phase structures, and many others.

This volume brings together work presented at the Fourth International Conference on Computational and Experimental Methods in Multiphase and Complex Flows. Featured topics include: Suspensions; Bubble and Drop Dynamics; Flow in Porous Media; Interfaces; Turbulent Flow; Injectors and Nozzles; Particle Image Velocimetry; Macroscale Constitutive Models; Large Eddy Simulation; Finite Volumes; Interface Tracking Methods; Biological Flows; Environmental Multiphase Flow; Phase Changes and Stochastic Modelling.

WIT Transactions on Engineering Sciences, Vol 56

ISBN: 978-1-84564-079-8   2007
apx 400pp
apx £130.00/US$235.00/€195.00

Atmosphere Ocean Interactions

Volume 2

Edited by: W. PERRIE, Bedford Institute of Oceanography, Canada

In recent years intense scientific research has been devoted to the understanding of atmosphere-ocean interactions. This volume continues and expands on some of the topics covered in Volume 1 (ISBN: 1-85312-892-9). The contributors consider hurricanes and severe marine storms within the context of atmosphere-ocean interactions. Factors related to storm intensity and evolution are mentioned, while impacts of storms on the upper ocean and related atmospheric variables are described.

Relationships between climate time-scales and storms and the impacts of storms on climate-related factors such as air-sea gas transfer are also considered.

Series: Advances in Fluid Mechanics Vol 39

ISBN: 1-85312-929-1   2006   240pp
£79.00/US$142.00/€118.50

WIT Press is a major publisher of engineering research. The company prides itself on producing books by leading researchers and scientists at the cutting edge of their specialities, thus enabling readers to remain at the forefront of scientific developments. Our list presently includes monographs, edited volumes, books on disk, and software in areas such as: Acoustics, Advanced Computing, Architecture and Structures, Biomedicine, Boundary Elements, Earthquake Engineering, Environmental Engineering, Fluid Mechanics, Fracture Mechanics, Heat Transfer, Marine and Offshore Engineering and Transport Engineering.

Vorticity and Turbulence Effects in Fluid Structure Interaction

An Application to Hydraulic Structure Design

F. TRIVELLATO, University of Trento, Italy

This book contains a collection of 11 research and review papers which have been contributed by each research unit joining the MIUR funded project: "Influence of vorticity and turbulence in interactions of water bodies with their boundary elements and effects on hydraulic design". The book features state-of-the-art Italian research devoted to the topic of fluid-structure interaction.

Series: Advances in Fluid Mechanics, Volume 45

ISBN: 1-84564-052-7 2006 304pp
£98.00/US$178.00/€147.00

Transport Properties of Organic Liquids

G. LATINI, Universita Politecnica delle Marche, Italy, R. COCCI GRIFONI, Universita delle Marche, Italy, G. PASSERINI, Universita Politecnica delle Marche, Italy

The liquid state is possibly the most difficult and intriguing state of matter to model. Organic liquids are required, mainly as working fluids, in almost all industrial activities and in most appliances (e.g. in air conditioning). Transport properties (namely dynamic viscosity and thermal conductivity) are possibly the most important properties for the design of devices and appliances. Most theoretical studies on the liquid state date back to the Fifties however huge advances in experimental studies and applied research on heat and mass transfer in liquids have been achieved during past decades. Most of the models cannot rely on theory alone and are empirical, while for most organic liquids, only a few experimental points and empirical correlations are available in literature.

The aim of this book is to present both theoretical approaches and the latest experimental advances on the issue, and to merge them into a wider approach. The book is organised into five chapters. The first chapter presents our theoretical knowledge of the liquid state. The second presents the tentative models for the evaluation of the thermal conductivity of organic liquids and confronts their results with the experimental data available in literature. The third presents the tentative models for the evaluation of the dynamic viscosity of organic liquids and confronts their results with the experimental data available in literature. The fourth presents a deeper review of the choice methods for thermal conductivity and their applications to mixtures of organic liquids and the fifth chapter presents a deeper review of the choice methods for dynamic viscosity and their applications to mixtures of organic liquids.

Series: Advances in Fluid Mechanics, Volume 46

ISBN: 1-84564-053-5 2006 208pp
£95.00/US$170.00/€142.50

Instability of Flows

Edited by: M. RAHMAN, DalTech, Dalhousie University, Canada

A state-of-the art analysis of studies in the field of instability of flows, this book contains chapters by leading experts in fluid mechanics. The text brings together many important aspects of flow instabilities and one of the primary aims of the contributors is to determine fruitful directions for future advanced studies and research.

Contents: Preface; Contact-Line Instabilities of Driven Liquid Films; Numerical Simulation of Three-Dimensional Bubble Oscillations; Stratified Shear Flow - Instability and Wave Radiation; Instability of Flows; Stability, Transition and Turbulence in Rotating Cavities; A Comprehensive Investigation of Hydrodynamic Instability.

Series: Advances in Fluid Mechanics, Vol 41

ISBN: 1-85312-785-X 2005 248pp
£89.00/US$142.00/€133.50

WIT eLibrary

Home of the Transactions of the Wessex Institute, the WIT electronic-library provides the international scientific community with immediate and permanent access to individual papers presented at WIT conferences. Visitors to the WIT eLibrary can freely browse and search abstracts of all papers in the collection before progressing to download their full text.

Visit the WIT eLibrary at
http://library.witpress.com

Numerical Models in Fluid-Structure Interaction

Edited by: S. K. CHAKRABARTI, Offshore Structure Analysis Inc., USA

This book covers a wide range of numerical computation techniques within the specialized area of fluid mechanics. Numerical computation methods on the effects of fluid on structures are described, with particular emphasis on the offshore application.

The book emphasizes the latest international research in the area for the advancement of the interaction problem and new applications of the development to the real world problems. The basic mathematical formulations of fluid structure interaction and their numerical modeling are discussed with reference to the physical modeling of the interaction problems.

The state-of-the-art on numerical methods in fluid-structure interaction is included, with emphasis on detailed numerical methods. Examples of the numerical methods and their validations and accuracy check are given, stressing the practical application of the problem. Some interesting results on numerical procedure are cited showing the limiting criteria of the numerical methods and typical execution time.

Series: Advances in Fluid Mechanics, Vol 42

ISBN: 1-85312-837-6 2005 448pp
£150.00/US$240.00/€225.00

All prices correct at time of going to press but subject to change.
WIT Press books are available through your bookseller or direct from the publisher.

Advances in Fluid Mechanics V

Edited by: C.A. BREBBIA, Wessex Institute of Technology, UK,
A.C. MENDES, University of Beira Interior, Portugal and M. RAHMAN, Dalhousie University, Canada

In this book highly regarded scientists, engineers and other professionals from around the world detail their latest research. Originally presented at the Fifth International Conference on Advances in Fluid Mechanics, the papers encompass a wide range of topics including: Fluid Structure Interactions; Hydrodynamics; Coastal and Estuarine Modelling; Bio-Fluid Mechanics; Boundary Layer Flows; Numerical and Experimental Methods in Fluids.

WIT Transactions on Engineering Sciences, Vol 45
ISBN: 1-85312-704-3 2004 472pp
£172.00/US$275.00/€258.00

Debris Flow

Phenomenology and Rheological Modelling

G. LORENZINI and N. MAZZA, University of Bologna, Italy

Debris flows are among the most frequent and destructive of all geomorphic processes and the damage they cause is often devastating. Increased anthropisation calls for improvements in the criteria used to identify debris-flow risk areas and the prevention measures adopted.
One of the main difficulties encountered by the approaches illustrated in previous literature is linked to their possible validation either in the field or in a laboratory environment.
The choice of a rheological model is extremely important. This book provides methodological details, which can be applied to investigations on debris-flow mechanics, capable of providing an accurate representation of the phenomenology.
Contents: Debris Flows - An Intermediate Phenomenon between Mass Wasting and Solid Transport; The Rheology of Hyperconcentrated Flows; Debris Flow Phenomenology; The State of the Art of Modelling Debris Flow Triggering and Mobilisation Mechanisms; The State of the Art of Modelling Debris Flow Dynamics of Motion; Debris Flow Modelling - A General Outline of the Motion Problem; Debris Flow Disaster Mitigation.

ISBN: 1-85312-802-3 2004 216pp
£80.00/US$128.00/€120.00

Nonlinear Flow Using Dual Reciprocity

W.F. FLOREZ, Pontifical Bolivariana University, Colombia
"...a valuable tool that should be present on the shelves of any beginner in the field."
APPLIED MECHANICS REVIEWS

Partial Contents: Integral Identities for Scalar and Vector Fields; Multi-Domain Formulation of the Navier-Stokes Equations; Multi-Domain Formulation of the Stokes Equations for Non-Newtonian Fluids; Numerical Results for the Navier-Stokes Equations.

Series: Topics in Engineering, Vol 38
ISBN: 1-85312-860-0 2002 148pp
£63.00/US$97.00/€94.50